DATE DUE

GAYLORD			PRINTED IN U.S.A.

AN INTRODUCTION TO METHODS & MODELS IN
Ecology, Evolution, & Conservation Biology

AN INTRODUCTION TO
METHODS & MODELS IN
Ecology, Evolution, &
Conservation Biology

STANTON BRAUDE &
BOBBI S. LOW

Editors

PRINCETON UNIVERSITY PRESS
PRINCETON AND OXFORD

Copyright 2010 © by Princeton University Press
Published by Princeton University Press, 41 William Street,
Princeton, New Jersey 08540

In the United Kingdom: Princeton University Press, 6 Oxford Street,
Woodstock, Oxfordshire OX20 1TW

Library of Congress Cataloging-in-Publication Data

An introduction to methods and models in ecology, evolution, and conservation biology /
Stanton Braude and Bobbi S. Low, editors.

p. cm.
Includes index.
ISBN 978-0-691-12723-1 (hardcover : alk. paper) — ISBN 978-0-691-12724-8 (pbk. : alk. paper)
1. Ecology—Research.   2. Evolution (Biology)—Research.   3. Conservation biology—
Research.   I. Braude, Stan.   II. Low, Bobbi S.
QH541.2.I65 2009
577.072—dc22   2009012206

British Library Cataloging-in-Publication Data is available

This book has been composed in Minion

Printed on acid-free paper.

press.princeton.edu

Printed in the United States of America

1 3 5 7 9 10 8 6 4 2

Contents

Figures

Tables

Many biology courses are offered with laboratory sections that teach the techniques specific to that discipline as well as the broader tools of how we do science. While this text cannot replace the hands-on experience of an ecology lab, it does introduce many of the theoretical and quantitative tools of ecology, conservation biology, and environmental science, and often shows how they intersect.

The exercises in this text were written and piloted by a group of teachers committed to helping students experience the intellectual excitement of ecology and environmental science, even when their courses may not give them the opportunity to gather their own data out in the field. These exercises have transformed our discussion sections into "brains-on" thinking labs rather than "hands-on" technique labs.

You will see that every exercise asks you not only to read, think, and "digest" the content, but also to analyze the information in specific ways, both alone (before class) and with others (in class). This is deliberate—we too have fallen asleep in class when all we had to do was listen! And we have assigned some of the most difficult tasks to be solved in small groups of students so that collaborative learning can take place.

You will also notice that we choose very simple techniques, often using paper and dice, for example, when there exist computer programs that can do the same task in a fraction of the time. This, too, is deliberate. For almost all of us, what is actually done in a computer is a mystery, a Black Box of methodology, if you will. We think it is *essential* to understand the process first, especially in simulation modeling. In part, you can explain better to others what you have done, if you have actually performed the process, rather than simply entering data. It is also true that if you understand the process thoroughly, you will be better at catching problems in later computer runs—you will have an intuition about the approximate answer, so that if you have mis-entered a data point (e.g., 20 rather than 2.0), you won't slavishly copy the computer's answer. And, finally, you will be better prepared to explain computer simulations to others.

One of our aims is to show how, even though we do not typically recognize it, ecology (section I), demography and population biology (section II), and population genetics (section III) are all closely related. Further, all these fields require that you be able to do some forms of quantitative analysis (section IV), and to synthesize what others have done leading to our present understanding, and to think about the current state of affairs (section V). It is not intended that any one course would use all of the chapters you find here. But the subset chapters used in different courses will overlap very differently depending on the approach and interests of your instructor. You may have this book as a supplemental text in more than one course—in fact, even if you do not, we hope that you will find some unassigned sections useful in other courses.

Acknowledgments

In addition to the contributing authors, a number of colleagues have piloted these exercises and provided valuable feedback: Nichole Bahls, Russell Blaine, Debbie Boege, Jason Bradford, Kuo-Fang Chung, Matt Gifford, Deena Goran, Rosie Koch, Allison Miller, Jennifer Neuwald, and Ty Vaughn. Their feedback has helped refine these exercises and make them even more valuable to students in their courses.

Special thanks go to James Robertson. In addition to authoring many of these chapters, James worked with many other contributors to design and develop their own chapters. He taught these exercises for five years; he is both an original biologist and a dedicated and creative teacher.

We would also like to thank Nancy Berg, Carl Simon, and Michael Low, who have read various chapters and provided extensive editorial help. Kate Malinowski taught herself *In-Design* and laid out the first draft of this text. Leah Corey, Carole Shadley, and Rebecca Martin very patiently compiled, edited, and formatted revisions.

We are also grateful to the Kemper Foundation, which supported the development of these exercises with a teaching innovation grant through Washington University.

Introduction

WELCOME TO *An Introduction to Methods and Models in Ecology, Evolution, and Conservation Biology*. We hope you will enjoy using it. The fields of ecology, evolution, behavior, and conservation, although treated as separate topics, in fact are aspects of a large interdisciplinary core of knowledge with a common theoretical foundation—we hope you will find that the skills you acquire are useful in many contexts. The best work in all of these fields begins with hypotheses about "how things work" and proceeds to devise experiments or collect data to test clear predictions that are derived from the hypotheses. The point, of course, is to devise tests so that the answers will distinguish among *alternative hypotheses*—different explanations that cannot simultaneously be true.

You will also find that we do something that may strike you as a step backward: we ask you to do a lot of pencil-and-paper work, plotting things as you think them through, for example. This is actually deliberate. We have found (as we bet you have, too) that it's altogether too easy to "cookbook" a process such as a statistical test without actually understanding just what we are doing. Only if you really understand just what each equation, each process *does,* will you be able to know when to use each in new situations, and how to apply each to new data.

Just how you use this text will depend on the particular course(s) in which you are using it, so you may not begin at the beginning, or go through the chapters in a linear fashion. In fact, if this text has been assigned in one of your courses, you may find it useful (we hope so) in others. Do, please, browse through!

Section I focuses on the foundations of evolutionary ecology: natural selection, adaptation, phylogeny, and life history analysis. In section II, we examine more traditional ecological models, from the Lotka-Volterra competition and predator/prey models to MacArthur and Wilson's island biogeography model. In section III, we deal with the basic population genetic parameters so frequently involved in making conservation decisions, but which are rarely well understood. You will use these to design conservation programs, for example. Section IV is a bit different. These chapters are organized around quantitative tools that we need to examine a wide array of ecological systems. You may find that you return to the statistics chapters for years, as you work to understand statistical language in scientific papers or when you choose statistical tests for your own independent projects. Finally, section V has synthetic exercises we hope will help you pull together a variety of skills you have learned this semester in the service of making broad applied or theoretical arguments.

Section I

Evolutionary Biology

1 Evolution and Pesticide Resistance: Examining Quantitative Trends Visually

Stanton Braude and John Gaskin

Introduction and Background

Evolution and natural selection have always been central concepts in the study of ecology. When German biologist Ernst Haekel coined the term "ecology" in the 1860s, he envisioned studying the forces of nature that were selective forces in the Darwinian sense. Darwin is popularly associated with the rise of evolutionary thought in biology; his major contribution was explaining natural selection—and the concept is so rich that we still find it fascinating to explore today.

Evolution is the term we use for changes in gene frequencies in populations or species over time. It is not the same as natural selection; in fact, evolution results from mutation, recombination, and drift, which generate variation but are not predictable, as well as from natural selection. So what is natural selection? It is the mechanism that drives adaptive evolution; the result of the simple fact that in any environment, depending on the conditions of that environment, some variants—individuals with specified genetic traits—survive and reproduce better than others. If we understand how any environment shapes traits, favoring some and disfavoring other individuals who possess those traits, we can predict how traits should match environmental conditions—and how populations will change over time. We will see this throughout this book, especially in this chapter, and in chapters 2, 4, 5, 18, and 19.

Ecology is a very empirical science, so it is not surprising that much ecology of the early twentieth century was descriptive. Ecologists today know that understanding natural selection and evolution is central to understanding important "why" hypotheses—especially today, when we humans change environments (and thus selective pressures) rapidly without necessarily understanding our impacts.

"Why" hypotheses can be of several sorts (Tinbergen, 1963). Hypotheses that explain why phenomena exist in nature are ultimate hypotheses, and those that explain how things work are proximate hypotheses. Both are important, but it is especially crucial not to confuse the two; it is confusing and wrong to offer a proximate answer to an ultimate question. For example: why do birds fly south for the winter? "Because individuals in this species in this region that migrate seasonally survive and reproduce better than those that do not" is an ultimate answer (and you can see all sorts of testable predictions: whether hummingbirds will migrate when seed-eating species will not; whether migration will be associated with seasonal changes, etc.). "Because changing day length causes shifting hormone levels" is a

TABLE 1.1.
Farm pesticide use in the United States 1964–1990 (million pounds of active ingredients).

Year	Herbicides	Insecticides	Other	Total
1964	76	143	72	291
1966	112	138	79	328
1971	207	127	130	464
1976	374	130	146	650
1982	451	71	30	552
1986	410	59	6	475
1987	365	57	7	429
1988	372	60	8	440
1989	394	61	8	463
1990	393	64	8	485

Source: U.S. Department of Agriculture.
Notes: For the years 1964, 1966, 1971, and 1976, estimates of pesticide use are for total use on all crops in the United States. The 1982 estimates are for major field and forage crops only and represent 33 major producing states, excluding California. The 1986–1990 estimates are for major U.S. field crops. Data for 1990 are projections.

proximate explanation: it tells how the changes are operationalized. Ultimate answers are always about differential survival and/or reproduction; there can be myriad proximate ways that responses are mediated. Depending on your question, you will be more interested in one or the other level.

Pesticide resistance is an example of evolution in action. Pesticide use, in both the United States and worldwide, has increased dramatically over the past 30 years (table 1.1) and farmers today have access to a diverse chemical arsenal to protect their crops (table 1.2). As a result, food productivity is higher now than at any other time in human history. But are there hidden costs, as a result of the selection our pesticides impose, and the resulting evolution of pest species? Have we had impacts we did not foresee?

Agricultural pesticides are typically applied broadly, so they are likely to affect unintended, or nontarget, species. These side effects can harm everything from arthropod predators (e.g., spiders and preying mantises), to birds and fish that feed on dead arthropods, to humans who use contaminated water. Pesticides can have secondary impacts: they can, for example, affect endangered species both directly and indirectly, leading to loss of biodiversity.

The effectiveness of any given pesticide rapidly decreases soon after its first use, regardless of the target pest or the pesticide. Although there are hopes that genetically engineered crops and their associated pesticides will avoid this trend, there are evolutionary reasons to doubt the success of any simple pest-elimination program. One reason that insect pests frequently bounce back in higher numbers after spraying is that insect predator populations are also reduced by insecticides. Predators obviously affect the death rate of prey, which means prey often experience intense mortality. Prey (food) populations affect the birth rate

TABLE 1.2.
Types of pesticide, volume of use in the United States, and major crop uses.

Pesticide	Type	Active ingredients (1,000 lbs/year)	Major crop uses
Acephate	Insecticide	1,900	Citrus
Azinphos-methyl (Guthion)	Insecticide	2,500	Peaches, pome fruits
Captan	Fungicide	10,000	Apples, peaches, almonds, seeds, other fruits and vegetables
Carbaryl (Sevin)	Insecticide	10,005	Citrus, fruit, nuts, fodder
Chlorothalonil (Bravo)	Fungicide	7,587	Fruits, vegetables, peanuts
Chlorpyrifos (Dursban/Lorsban)	Insecticide	7,023	Citrus, corn, fruit, grain, nuts, vegetables
Demeton	Insecticide	165	Vegetables and orchard crops
Diazinon (special review)	Insecticide	2,125	Fruits, nuts, livestock, lawn and turf
Dicloran (DCNA)	Fungicide	355	Peaches, plums, cherries, grapes, other fruits and vegetables
Dimethoate	Insecticide	1,453	Citrus, pome fruit, nuts, grapes, tomatoes, and many vegetables
Disulfoton	Insecticide	2,111	Grains, strawberries and pineapples, vegetables
Folpet	Fungicide	1,500	Grapes, apples, melons
Malathion	Insecticide	15–20 million	Many fruits and vegetables, tree nuts, grains, fodder
Mancozeb	Fungicide	16,000	Apples, onions, potatoes, tomatoes, small grains
Methamidophos	Insecticide	1,259	Potatoes, cotton, cabbage and other crops
Methyl parathion	Insecticide	8,934	Grains, peanuts, berries, many fruits and vegetables
Mevinphos	Insecticide	1,278	Many vegetables and fruits
Monocrotophos	Insecticide	760	Peanuts, sugarcane, tobacco, potatoes, tomatoes
Omethoate (Folimat)	Insecticide		Fruit crops, vegetables, hops
Parathion	Insecticide	7,000	Citrus, cotton, orchard crops, vegetables, fruits
Permethrin (Ambush/Pounce)	Insecticide	1,475	Vegetables
Quintozene (PCNB)	Fungicide	2,523	Vegetables, small grains

Note: This list is for pesticides studied in the Natural Resources Defense Council study, *Intolerable Risk* (1989), and is drawn chiefly from a National Research Council report, *Regulating Pesticides in Food* (1987).

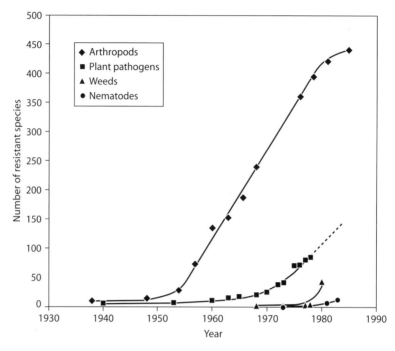

FIGURE 1.1. The number of pesticide-resistant species has grown exponentially since the start of widespread use of insecticides and herbicides in the middle of the last century (modified from National Research Council Report, 1986).

of predators, which means prey (in this case pest) populations recover more rapidly than predators (this phenomenon is called Volterra's principle). This is hardly a desired result, because we lose the natural predators of the pests.

The main cause of a decrease in pesticide effectiveness results from the evolution of resistance to the pesticide as a result of natural selection (figure 1.1). Think of it this way: if a pesticide is 95% effective, it kills 95% of the pest individuals—but the remaining individuals are the resistant ones, and their progeny will be more resistant, on average, than individuals in the parent population. We call this directional selection—selection that favors one extreme (here, pesticide resistance).

Thinking about pesticide use in agriculture (or widespread antibiotic use in medicine) raises some interesting evolutionary questions. Many of our newly developed pesticides and herbicides have natural analogs. In this case, because our target pest species may have had long exposure to the natural compounds, we are not surprised if resistance has evolved in species in which there were alleles that conferred resistance. Resistance-conferring alleles can persist for various reasons, even before we begin treatment. Remember that variation (e.g., in resistance) arises from mutation, drift, and recombination. It also matters whether resistance is expensive. The level of existing resistance ability thus depends on several things: frequency of exposure, cost of developing resistance, cost of being nonresistant, and the generation time. The patterns you see in this chapter are so strong because high initial death rates (see tables below) mean that the cost of not being resistant is very high; and most pests have a fairly short life span (many generations = many chances for variations to arise, and selection to act), so that evolution proceeds quickly.

Objectives of this Exercise

In this exercise you will:

- Examine two case studies of pesticide use and effectiveness
- Evaluate the effect of pesticides as a selective agent driving evolution
- Plot the data to help you identify trends more easily

Case Studies and Data

Read the following two cases, examine the data, and answer questions 1 through 7.

Cotton

Almost 50% of all insecticides applied to crops in the United States are applied to cotton! As a result, most major insect pests of cotton have developed resistance to one or more of these insecticides. Some cotton pests, such as the tobacco budworm (*Heliothis virescens*) and spider mites (*Tetranychus* species), are now resistant to most of the insecticides registered for use on cotton in the United States. Our arsenal of effective insecticides for use on cotton is rapidly disappearing. Further, many insecticides are indiscriminate killers, destroying the predatory arthropods (e.g. mud wasps, ladybird beetles, dragonflies) that normally control insect herbivore populations, and giving rise to the problem outlined above.

In the 1930s, vast crops of cotton were grown in southern Texas and northeastern Mexico. Boll weevils, pink bollworms, and cotton flea hoppers were the key pests of the crop. These pests were controlled with calcium arsenate and sulfur dust, which quickly yielded profitable crop harvests. There were sporadic, but not devastating, outbreaks of another cotton pest, the tobacco budworm. Shortly after World War II, new chlorinated hydrocarbon insecticides, such as DDT, became available. These pesticides provided greater crop yields, because they destroyed almost all cotton pests. Some pesticide regimes called for 10–20 DDT applications per growing season, but the future of cotton in southern Texas was looking great!

Unfortunately, these pesticide treatments also decimated the predators who feed on the pest species. Because this led to occasional increases in pest numbers, the dosage of pesticide applications was increased. In the mid-1950s the boll weevil developed resistance to chlorinated hydrocarbon insecticides. By 1960, the pink bollworm and the tobacco budworm had also become difficult to control (even with treatments of 1–2 lb. per acre every 2 days!). By 1965, all of the cotton pests were resistant to DDT and similar pesticides. An organophosphorus pesticide (methyl parathion) was called into use, but by 1968 the tobacco budworm had developed resistance to methyl parathion as well. Cotton farming in northeastern Mexico was reduced 70-fold, from 70,000 acres in the 1960s to less than 1000 acres in 1970. These are examples of two unintended consequences of pesticide application: (1) increasing pesticide resistance through directional selection, and (2) decimating natural predators of the pest species. Table 1.3 shows data for this problem in southern Texas.

Apples

Phytophagous (plant-eating) mites such as *Tetranychus mcdanieli* [*T. mc*] and *Panonychus ulmi* [*P. ul*] are serious pests of apples in Washington state. The practice of preventive

TABLE 1.3.
Tobacco budworm (*Heliothis virescens*) resistance to methyl parathion in the lower Rio Grande Valley (RGV) and near College Station (CS), Texas, between 1967 and 1971.

Case	Year (location)	Methyl parathion lb./acre	% of tobacco budworm killed
A	1967 (CS)	0.5	91
		1.0	99
B	1968 (CS)	0.5	41
		0.75	58
		1.0	71
		1.5	90
		2.0	92
		2.5	91
		3.0	90
C	1968 (RGV)	0.5	30
		0.75	32
		1.0	50
		1.5	64
		2.0	71
		2.5	81
		3.0	85
D	1970 (CS)	1.0	21
		1.5	23
		2.0	42
		2.5	46
		3.0	50
E	1970 (RGV)	1.0	13
		1.5	17
		2.0	11
		3.0	31
F	1971 (CS)	1.0	0
		2.0	18
		3.0	12

scheduling—pesticides applied whether there was evidence of current pest problems or not—of pesticide application in the past has resulted in eradication of the natural enemies of the mites (which often included other mites such as *Metaseiulus occidentalis* [*M. oc.*]). To avoid these difficulties, integrated pest management programs, involving natural predators or a combination of selective pesticides and predators, have been developed to preserve the advantages of natural pest control in artificial ecosystems. Table 1.4 contains data on mite populations in orchards with different treatments.

Questions to Work on Individually Outside of Class

When farmers began using organic pesticides half a century ago, the hope was that we could eradicate pests and not have to share our yields with them. However, as you have seen in

TABLE 1.4.

Average number of two phytophagous (plant-eating) mites (*Tetranychus mcdanieli* [*T. mc*] and *Panonychus ulmi* [*P. ul*]) and one predatory mite (*Metaseiulus occidentalis* [*M. oc.*]) in an apple orchard.

Mite	Date	5/1	5/15	6/1	6/15	7/1	7/15	8/1	8/15	9/1	9/15
				Average number of mites per leaf							
				With a standard chemical spray program							
T. mc.			4.5	2.9	1.2	7.5	20	0.4	3.9	27	44	20
P. ul.			0.7	0.1	4.9	5.0	43	1.4	1.9	0.8	0.3	1.1
M. oc.			0.4	0	0	0.1	0.1	0.2	0	0	1.1	1.8
Mite	Date	5/1	5/15	6/1	6/15	7/1	7/15	8/1	8/15	9/1	9/15
				Unsprayed apple orchard							
T. mc.		250	80	0	0	0	0	0	0	0	0
P. ul.		0	6.0	4.8	5.5	5.0	0	0	0	2.0	0
M. oc.		0.5	2.7	2.9	2.5	2.0	0.7	0.7	0	0.3	0

Source: Hoyt, 1969. Integrated chemical control of insects and biological control of mites on apple in Washington. *Journal of Economic Entomology* 62: 74–86.

the data above, the pests are still there after repeated treatment. So, what is the effect of these pesticides on their intended targets? You should be able to see the effect as you work through the following questions.

1. Describe the effect of methyl parathion dose on budworm kill.

2. Graph the relationship between methyl parathion dose and percent budworm kill. All six cases can be plotted on a single graph, but each of the cases should be plotted as a separate line. Now look back at question 1 and see if you want to revise your answer.

3. How does the density of phytophagous mites change over time for both species, and for both treatments? How does the density of carnivorous mites change over time?

4. Graph the relationship between mite density and time for all three species, and for both treatments. (All six of these lines can be placed on one graph.) Now look back at question 3, and see how you can revise it to make it clearer with reference to your plot.

5. How do insecticides appear to affect the evolution of insect populations? Support your answer with reference to the data and the trends apparent in your graphs. Are there any other data you would want, in order to show that the change was the result of selection?

6. Population growth (whether we are talking about insects, plants, fish, or any species) depends on the balance among birth, immigration, death, and emigration.

(a) When you consider the growth of a prey population, which of these population parameters will be most affected by their interaction with predators?

(b) When you consider the growth of predator populations, which of these population parameters will be most affected by the size of the prey base?

(c) If a very cold winter wiped out 99% of both the predators and prey in a community one year, how do you expect this to affect the growth of the prey population the following spring? How do you expect it to affect the growth of the predator population the following spring?

7. In the apple mite example, the prey and predator had about the same generation time. This is not always the case: normally, the predator's generation time is longer than that of the prey—which slows down the predator's recovery after spraying even further than the fact that predator birth rates are affected by prey abundance. How could this difference in generation time affect population numbers over time if both predator and prey are present in the orchard at the time of spraying and both are initially susceptible to the chemical sprayed?

Small-Group/In-Class Exercise

When you come to class you may be asked to work on one or both of the following exercises. Bring five pieces of blank 8.5 × 11 paper for the Exercise A option.

Exercise A

In this exercise you will be asked by your teacher to copy a series of diagrams of mites. Follow his/her instructions. The discussion that follows will address concepts of evolution, natural selection, and drift.

Exercise B

In this exercise your group will have approximately one-half hour to outline an argument for or against the Mango Marketing Manifesto (the details are fictional, but your arguments should be based on a real understanding of evolution in response to pesticides). Your group may be asked to take the position expected of one of the following groups: Sierra Club, Monsanto Corporation, Local Farm Cooperative, Center for the Study of Intelligent Design and Creation, Anglers and Fly Fishermen of America, Peach Farmers and Orchardists of America.

Each group will have a few minutes to present their position to the class and you will have time to question each group after their presentation. Try to stay in character, but be sure to cite examples and data where relevant.

Congratulations! Your enrollment in this course has landed you a summer internship in the state capital. Each group of you will represent a different constituency, following your instructor's suggestions. Your first assignment is to review the following proposal, and comment on its scientific merit from your group's perspective. Your group has 30 minutes to discuss the proposal and to organize your presentation. You then have to make a concise five-minute report to the state agriculture committee.

Your predecessor submitted a report that this proposal is based on sound scientific reasoning. He argued that it is similar to the use of antibiotics in medicine. When someone gets a bacterial infection we give a course of antibiotics that totally exterminates the bacterial population in that individual, but we don't give antibiotics to every person in a population.

1. Should we support the proposal or not?
2. Should we make the argument that this is just like an infection and that similar treatment will be a good idea?
3. What are the effects at the population level (over time) of such treatments?
4. Explain why this plan will or will not work.

Box 1.1 The Mango Marketing Manifesto

In 1991 our State Agriculture Department reported that a delicious Nepalese variety of Mango can grow in our midwestern climate. This variety fetches premium prices in the fruit markets of Manhattan, Boca Raton, Santa Monica, and Winnetka. Our State Department of Land and Water subsidized the planting of Nepalese Mango groves throughout St. Joseph's and Loyola Counties. Unfortunately, the fruit fly, *Drosophila columbii*, attacks the ripe fruit and can reduce yields by over 20 percent.

The representative for St. Joseph's County proposes that we allocate $64,000 to support efforts to control this pest. Since much of the market for mangos is in the organic and health food sector of the economy, we propose that only 29 of the 87 mango growers in the bicounty area spray their groves intensively each year. They assure us that this will destroy the pest population in three years, and the unsprayed groves can still sell their produce in those swanky organic markets in two out of the next three years.

If this plan will not work, propose an alternative solution to the pest problem. Explain your reasoning and convince your audience with any and all evidence at your disposal (whether they are data you have worked with, or other available data).

Your team will have five minutes to present your argument to the committee, and will then take questions from them. (*Note*: In preparing for critical questions, you may also find questions that you should pose to other groups.)

References

Adkisson, P. L. 1982. Controlling cotton's insect pests: a new system. *Science* 216: 19–22.

Cox, G. 1993. *Conservation Ecology*. Boston, Mass.: Wm. Brown.

Hoyt, S. C. 1969. Integrated chemical control of insects and biological control of mites on apple in Washington. *J. Economic Entomology* 62: 74–86.

Hoyt, S. C. 1969. Population studies of five mite species on apple in Washington. *Proceedings of the Second International Congress of Acarology, Sutton Bonington, England (1967)*: 117–133. Budapest: Acad. Kiado.

Tinbergen, N. 1963. On aims and methods of ethology. *Zeitschrift fur Tierpsychologie* 20: 410–433.

World Resources Institute. 1992. *Environmental Almanac*. Boston, Mass.: Houghton Mifflin.

Young, H., and T. Young. 2003. A hands-on exercise to demonstrate evolution by natural selection and genetic drift. *The American Biology Teacher* 65: 444–448.

2 Lizard Ecomorphology: Generating and Testing Hypotheses of Adaptation

Kenneth H. Kozak

Introduction and Background

From at least Aristotle's time, naturalists and philosophers have commented that many organisms seem wellsuited to their environments. But Charles Darwin introduced a really novel and exciting twist to our view of this relationship. He began by reviewing what was known at the time (1850) about artificial selection: how we humans have shaped dog breeds, plants we are interested in, and more, by allowing some individuals (those with the traits we liked) to survive and breed, and prohibiting others. He then proposed that natural conditions also might impose selection on organisms. This natural selection, through the differential survival and reproduction of individuals with some traits (characteristics), would lead to the sorts of trait-environment "fit" we see. In the more than 150 years since Darwin, we have learned about the role of genetics, and we have expanded and refined our understanding of evolution, natural selection, and adaptation.

Evolution is the process of change in genetic composition of populations over time. Nothing will change, of course, unless some variation exists; much variation is generated by mutations—actual changes in the genetic structure that result from mistakes in replication of DNA. These happen often as a result of environmental insults such as UV radiation. Recombination of alleles in reproduction can also produce new combinations of alleles. Random changes in allele frequency due to accidental survival, reproduction, or dispersal can change gene frequencies; these are referred to as drift. Natural selection is like the "filtering" effect imposed by environmental conditions, because typically, in any environment, not all variants survive and reproduce equally well. So if we look at a hot, dry, environment like the Kalahari, we would expect to find organisms that tolerate or avoid heat (perhaps by being active only at night).

But there are two important caveats. First, because some trait would be advantageous does not mean that an organism will have that trait—this depends on whether the genetic variation exists for that trait to spread due to selection. Second, just because a trait looks handy does not mean that it is an adaptation. As the biologist George Williams pointed out long ago (1966), the concept of adaptation is an onerous one: you must show that the trait developed as a result of natural selection, not simply that the trait is advantageous. Richard Lewontin and Richard Levins proposed terms for some of the "handy-but-not-an-adaptation" conditions. Imagine something that does not enhance fitness but is not costly, so it continues to exist (if it were costly, selection would weed it out, like functional eyes in cave fish). These are not adaptations. Or consider something like the nasty skin secretions of toads. The original function of these secretions was to get rid of metabolic by-products in relatively dry environments (toads can live in drier environments than most frogs, in part because of this). But because of the chemistry of these secretions, they are really discouraging to predators. Now, that is certainly advantageous—but it was not the original function of the secretion. Skin secretion is thus not an adaptation for predator avoidance (though, when it confers an advantage in survival and/or reproduction we still call it adaptive); we can call it an exaptation.

We often have questions about whether a trait is an adaptation or not. We generate alternative hypotheses about what we should see if something is, or is not, a true adaptation, shaped by natural selection for a particular function (alternative hypotheses cannot both, or all, be true—only one can prevail). Strong support for the hypothesis that a trait is an adaptation requires multiple sources of evidence, including:

1. *Common ancestry*—we want to be able to trace the trait and its alternative states through a sequence of ancestors along an evolutionary tree (and we will see how to generate these in chapter 3).
2. *Correlation between a trait and an environmental condition*—if our definition above is right, we should see correlations between environmental conditions (heat, cold, range of variation in those, etc.) and traits evolved in response (heat tolerance, cold tolerance, etc.).
3. *Current utility of a trait*—simple correlation is not enough. We must be able to establish that the trait we are proposing as an adaptation actually confers a fitness advantage, in enhanced survival and/or reproduction, compared to competing traits. (Note that this can be difficult in rapidly changing environments.)

In this exercise you will examine apparently unconnected data to test hypotheses of adaptation in lizards. Keep in mind also that new natural history information may come to light and change the hypothesized status of a trait, for example from an exaptation to an adaptation.

Homework for this exercise takes approximately 30 minutes.

Objectives of This Exercise

In this exercise you will be provided with experimental, ecological, and evolutionary data on trait variation in several organisms. You will synthesize these diverse data and test the hypothesis that a trait is an adaptation.

Case Study and Data

Biologists have long been intrigued by the subdigital toepads that allow some lizards to cling to smooth surfaces (figure 2.1). Anoline lizards (figure 2.2), one of these groups, are associated with arboreal habitats in tropics of mainland South America, and the Greater and Lesser Antillean Islands of the Caribbean. Locomotion in arboreal habitats is achieved in two different ways: grasping and adhesion. Anoline toepads are modified and expanded scales called lamellae. The lamellae are covered with millions of microscopic setae, tiny hairlike structures. These setae form bonds with electrons on the substrate, and facilitate adhesion to smooth surfaces. This is why anoles can scamper up glass windows. In theory, lizard claws could provide some grasping ability, but one major hypothesis is that toepads provide both adhesion and improved grasping ability.

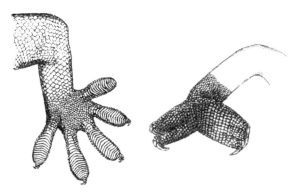

FIGURE 2.1. Like anoles, geckos have foot pads (left) but there are other means of clinging to a branch. For example, chameleons have opposing digits with long claws (right).

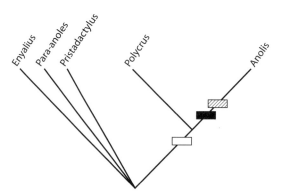

FIGURE 2.2. Anole lizard.

As an evolutionary ecologist you are interested in testing the hypothesis that the toepads of anoline lizards are an adaptation that evolved under natural selection for the purpose of efficient arboreal locomotion. One line of evidence regarding whether a trait is an adaptation depends on when it arose in evolutionary time. Fortunately, a systematist has generated a hypothesis of the evolutionary relationships of anoline lizards and their close relatives. With data on the ecology of anoline relatives we can estimate when the ancestors of anoles first invaded arboreal habitats. We can also map the first appearance of toepads in the evolutionary history of anoles (figure 2.3).

FIGURE 2.3. An evolutionary tree depicting the evolution of toepads in anoline lizards. The open bar indicates the origin of arboreality in an ancestor of *Polycrus* and *Anolis*. The black bar indicates the ability to cling to smooth surfaces, and the hatched bar indicates the origin of toepads. These characteristics arose in the common ancestor of *Anolis*, but there is no evidence which came first.

Another line of evidence regarding whether a trait is an adaptation would be that the trait has current utility. A laboratory experiment tested the clinging ability of lizards with, and without, toepads. A sample of *Anolis* lizards (all of which have toepads) and related taxa that lack toepads were placed on a nearly vertical (85° angle) smooth plate in the laboratory. Each individual lizard was then pulled off the plate four times. The force required to remove each lizard was recorded. Examine table 2.1, which illustrates the data from these experiments.

TABLE 2.1.
Body mass and clinging ability for fifteen species of anoline lizards.

Species	Mass (g)	Clinging ability (N)
Anolis cristatellus	8.1 ± 0.26	2.837 ± 0.335
A. cuvieri	44.5 ± 3.98	14.857 ± 0.485
A. evermanni	5.6 ± 0.28	3.148 ± 0.223
A. gundlachi	7.1 ± 0.18	2.305 ± 0.228
A. krugi	2.4 ± 0.09	2.162 ± 0.122
A. occultus	0.5 ± 0.04	0.834 ± 0.159
A. poncecsis	1.6 ± 0.08	1.312 ± 0.124
A. pulchellus	1.5 ± 0.04	1.463 ± 0.147
A. stratulus	1.9 ± 0.10	1.471 ± 0.119
A. garmani	31.7 ± 1.50	13.484 ± 1.728
A. grahami	6.2 ± 0.23	2.344 ± 0.200
A. lineatopus	4.6 ± 0.51	1.876 ± 0.244
A. opalinus	2.1 ± 0.08	1.275 ± 0.165
A. sagrei	2.9 ± 0.09	1.136 ± 0.132
A. valencienni	6.8 ± 0.73	3.102 ± 0.210
Polycrus	—	Untested
Enyalius	—	0
Pristadactylus	—	0
Para-anoles	—	0

Note: The measurements shown are the mean and standard deviations for a minimum of 15 individuals for each species. Lizards lacking toepads (*Enyalius*, *Pristadactylus*, and para-anoles) could not cling to the smooth surface. Data are unavailable for *Polycrus*.

Questions to Work on Individually Outside of Class

1. How does the basic natural history of lizards point to the hypothesis that toepads are an adaptation? For what would toepads be an adaptation? Where do these animals live?

2. Which data helped you discriminate between a nonaptative and an aptative role for toepads?

3. In what ways might the increased clinging ability provided by toepads be important to the fitness of these lizards? What selective advantage might toepads confer?

4. What other information do you need to test the hypothesis that toepads are actually an adaptation?

Small-Group / In-Class Exercise

When you come to class all groups will be asked to work on the following four problems. Once you have completed that work, one group will present their findings and conclusions in the format of a scientific seminar. The other groups will be assigned roles as audience members.

(a) Some groups will act as students of Richard Lewontin or Stephen Jay Gould (staunch critics of adaptationism).
(b) Some groups will act as students of Richard Dawkins (proud adaptationist).
(c) Some groups will act as reviewers for the journal *Animal Behaviour* (careful critics of any argument).

After the presentation, each group will have an opportunity to question the presenters from the point of view their character is likely to take.

First, work as a group to answer these questions about the links between ecology and morphology in Caribbean Anolis lizards.

Supporting Hypotheses of Adaptation with Multiple Lines of Evidence

For a trait to be an adaptation it must be shaped by evolution to perform the adaptive function. Strong support for the hypothesis that a trait, such as relative limb length, is an adaptation for locomotion in a particular habitat would require various types of evidence including: common ancestry, correlation between the trait and the environment, and current utility of the trait. Your group may be asked to make the case that anole ecomorphs are adaptations to the different habitats in which they are found (summarized in table 2.2). Working through the following problems may help you support the hypothesis of adaptation. But you are as likely to have the job of critiquing that hypothesis, and understanding these lines of inference will help you understand where the case is weakest.

Problem I: Ecological and Morphological Diversity (Questions 5–6)

Anolis exhibits remarkable species diversity. Nearly 300 species have been described; about half of these occur on Caribbean islands. In the Greater Antilles (Cuba, Hispaniola, Jamaica, and Puerto Rico) these lizards occur in assemblages of species that differ in the habitats that they use. As many as six different ecomorphs—species that inhabit particular microhabitats—may be found in sympatry. These six ecomorphs are named according to the microhabitat where they are most frequently found: grass-bush, trunk-ground, trunk, trunk-crown, crown giant, and twig dwarf (table 2.2). Each of the ecomorphs differs in its perch use, morphology, and behavior (tables 2.2 and 2.3).

5. We are interested in the relationship between the morphology of "legginess" and perch diameter for each of the species. However, limb length alone would be a confusing measure

Table 2.2.
Morphology, perch diameter, and sprint performance of the six *Anolis* ecomorphs.

Feature	Ecomorph					
	Crown giant	Twig dwarf	Trunk-crown	Trunk	Trunk-ground	Grass-bush
Size	>100 mm	<50 mm	>70 mm	>60 mm	>60 mm	<50 mm
Modal perch	Typically high in tree crown	Twigs of canopy	Canopy and upper trunk	On trunk between trunk-crown and trunk-ground	On lower trunk	On grasses or bushes
Body proportions	Head large, massive, often casqued	Long head, short body, short legs	Large head, body tending to be long, short legs	Head and body short	Head relatively short, body short and stocky, limbs long	Head moderately long, body slender, tail long
Foraging behavior	Primarily a canopy forager	A slow searcher on twigs	A searcher on leaves and branches	Primarily a forager on its trunk perch	Sit-and-wait predator on ground prey	Primarily a grass-bush forager
Defense behavior	Primarily aggressive	Primarily cryptic	Flight upward	Squirreling (rapid movement to the opposite side of the trunk)	Flight downward	Flight downward

Source: Williams, 1983.

because bigger animals have longer limbs. To correct for this, we can adjust limb length for overall animal size. Does there seem to be a relationship between size-adjusted limb length and perch diameter? (You need to plot the relationship to answer this question.) Because there is such a wide range of perch diameters, perhaps the relationship will become clearer by plotting size-adjusted limb length against the natural logarithm (ln) of perch diameter.

6. Do relationships among hindleg length, perch diameter, and the ecomorph categorizations exist? If so, describe the relationship(s)?

Problem II: Morphology, Perch Diameter, and Sprint Performance (Questions 7–9)

In nature, correlations between organism morphology and ecology often exist. Take the finches of the Galapagos Islands for example. In these species, bill morphology reflects the size and hardness of the seeds that are most often eaten (Grant, 1981). In the deserts of Australia, skinks with long legs use more open habitats, whereas those with short legs are more frequently associated with dense vegetation (Pianka, 1969). Strong associations between traits and environments, such as these, shaped Darwin's theory of organismal change by natural selection. If natural selection has favored a particular trait in a specific environment, it follows that variation in the trait should be directly related to the performance capabilities of the organisms in the context of interest. Here we are interested in the connection between the locomotory abilities of various lizard morphs in different parts of the forest canopy.

A laboratory experiment was carried out to evaluate the effect that perch diameter has on sprint speed in eight species of *Anolis* that differ in hind-leg length. Maximum sprint speed was measured on wooden dowels of varying diameters to simulate a variety

TABLE 2.3.
Morphological and ecological measurements for seventeen species of *Anolis*.

Species	Morphological measurements			Ecological measurements		
	Snout-vent length (mm)	Mass (g)	Hindleg length (mm)	Perch height (m)	Perch diameter (cm)	Ecomorph type
angusticeps	—	—	—	—	—	Trunk-ground
carolinensis	—	—	—	—	—	Trunk-crown
cristatellus	63.8	8.1	53.8	1.2	13.3	Trunk-ground
cuvieri	127.0	44.5	95.5	—	—	Crown giant
evermanni	62.3	5.6	47.9	3.4	35.3	Trunk-crown
gundlachi	65.1	7.1	57.6	1.3	33.5	Trunk-ground
krugi	48.6	2.4	38.9	0.6	5.7	Grass-bush
occultus	38.1	0.5	16.5	—	—	Twig dwarf
poncensis	43.9	1.6	31.2	0.9	3.6	Grass-bush
pulchellus	43.6	1.5	32.5	0.3	1.4	Grass-bush
stratulus	44.5	1.9	32.8	7.0	9.4	Trunk-crown
garmani	109.3	31.7	80.0	3.5	33.2	Crown giant
grahami	61.6	6.2	45.6	2.5	17.7	Trunk-crown
lineatopus	57.2	4.6	46.1	1.0	26.0	Grass-bush
opalinus	47.7	2.1	34.1	1.4	24.2	Trunk-crown
sagrei	48.8	2.9	36.2	0.4	15.4	Trunk-ground
valencienni	72.1	6.8	38.5	2.3	6.0	Twig dwarf

of branch sizes that these lizards encounter in nature (0.7, 1.6, 2.5, and 5.1 cm in one series of trials and 1.2, 2.1, 2.6, 3.3, and 4.6 cm in a second series of trials). The dowels were angled at 37° and 45° in each of the two series of trials, respectively. Lizards were tested on each size rod once per day, and this sequence of trials was repeated on four separate occasions. The results of these speed tests are presented in figure 2.4. However, we are not just interested in how fast the lizards can run but in the connection between their running ability and the environment. In order to measure this, Irshick and Losos (1999) created an index of sprint sensitivity. This is a measure of how much perch diameter affects the sprinting ability of a lizard. Table 2.4 tells us how species differ from one another in their sprint sensitivity.

7. In general, how do the lizard's sprinting capabilities respond to branches with smaller diameters?

8. In what species is sprinting ability least affected across the range of branch diameters? To which ecomorph categories do the less sensitive species belong? Do any general trends emerge with respect to the different ecomorphs and their sprinting abilities?

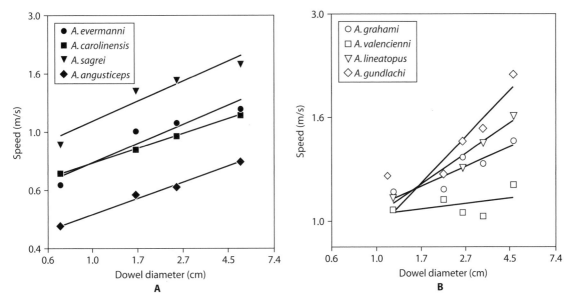

FIGURE 2.4. All eight species of *Anolis* can run faster on larger-diameter dowels than on smaller-diameter dowels (and hence run faster on larger branches). However, some species are especially slow on small branches (Irshick and Losos, 1999).

TABLE 2.4.
Comparisons of sprint sensitivities for eight species of *Anolis* over the range of dowel diameters.

Species	AS	AL	AGu	AC	AE	AGr	AA	AV
AS								
AL	ns							
AGu	ns	ns						
AC	ns	ns	*					
AE	ns	ns	ns	ns				
AGr	ns	ns	*	ns	ns			
AA	ns	ns	*	ns	ns	ns		
AV	*	*	**	*	*	ns	*	

Source: Irshick and Losos (1999).

Notes: AS = *Anolis sagrei*, AL = *A. lineatopus*, AGu = *A. gundlachi*, AC = *A. carolinensis*, AE = *A. evermanni*, AGr = *A. grahami*, AA = *A. angusticeps*, and AV = *A. valencienni*.

ns denotes that the two species did not differ in their sprint sensitivities over the range of dowel diameters

* and ** denote that the two species differed significantly in their sprint sensitivities at $P < 0.05$, and $P < 0.01$ alpha levels.

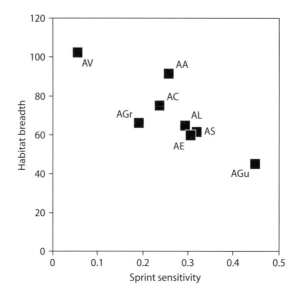

FIGURE 2.5. *Anolis* habitat generalists tend to be less sensitive to branch diameter than specialists.

9. Given what is known about the defense and predatory behaviors of the different eco-morphs (see table 2.2), formulate a hypothesis to explain why some species always sprint faster than others across the range of branch sizes.

Problem III: Sprinting Capabilities and Habitat Selection

While laboratory performance studies provide a measure of performance capability across a range of conditions and may provide clues about the utility of a given trait, an important question remains: are the performance measures ecologically relevant? For example, it is important to know whether the species evaluated in performance experiments actually utilizes the conditions in which it performs best in its natural habitat. Thus, the ability of an organism to select preferentially the habitats in which it excels plays a very important role in enhancing fitness.

Consider the following field study of the eight species of *Anolis* evaluated for their sprinting capabilities. Each of the species was videotaped in its natural habitat. Careful observation during playback allowed researchers to calculate the range of perch diameters in the wild during the observation period. The relationship between the range of perch diameters (perch habitat breadth) and sprint sensitivity was explored for eight different *Anolis* species (figure 2.5).

10. Which species used the largest and smallest range of habitats, respectively?

11. What is the relationship between sprint sensitivity and the range of perch diameters used by these species in their natural environment?

12. What do the field and lab performance data suggest about the structural habitats these species utilize in their natural environment? Do species prefer the branch diameter on which they are the fastest?

13. Are the laboratory and field data sufficient to support the hypothesis that relative hind-leg length is an adaptation for locomotion in different parts of the forest canopy? Briefly summarize your reasoning.

Problem IV: Ecomorph Evolution (Questions 14–15)

An amazing aspect of anole diversity is that the same set of ecomorphs is present on different islands in the Caribbean. Given that dispersal between islands is an unlikely scenario, early investigators proposed that the same set of ecomorphs has evolved independently on each of the Greater Antillean islands. Recent phylogenetic research on the evolutionary relationships of *Anolis* has confirmed the hypothesis of within-island ecomorph radiations. In other words, Cuban trunk-ground anole species are more closely related to the other Cuban anole species than they are to trunk-ground anole species on Jamaica or to trunk-ground anole species on other islands.

14. What do these data on ecomorph evolution suggest regarding the selective environments on each of the Greater Antillean islands?

15. Given what you know about the ecology, morphology, behavior, performance capabilities, and evolution of Caribbean *Anolis* lizards, do you consider hind-leg length an adaptation?

References

Irschick, D. J., and J. B. Losos. 1999. Do lizards avoid habitats in which performance is submaximal? The relationship between sprinting capabilities and structural habitat use in Caribbean anoles. *American Naturalist* 154: 293–305.

Gould, S. J., and E. S. Vrba. 1982. Exaptation—a missing term in the science of form. *Paleobiology* 8: 4–15.

Grant, P. R. 1981. The feeding of Darwin's finches on *Tribulus cistoides* (L) seeds. *Animal Behaviour* 29: 785–793.

Levins, R., and R. Lewontin. 1985. *The Dialectical Biologist*. Cambridge, Mass.: Harvard University Press.

Pianka, E. R. 1969. Sympatry of desert lizards (*Ctenotus*) in Western Australia. *Ecology* 50: 1012–1030.

Williams, E. E. 1983. Ecomorphs, faunas, island size and diverse endpoints in island radiations of Anolis. Pages 326–370 in R. B. Huey, E. R. Pianka, and T. W. Schoener (eds.) *Lizard Ecology: Studies of a Model Organism*. Cambridge, Mass.: Harvard University Press.

Williams, G. C. 1966. *Adaptation and Natural Selection*. Princeton, N.J., Princeton University Press.

3
Phylogenetic Inference: Examining Morphological and Molecular Datasets
James Beck

Introduction and Background

To understand organisms and their interactions we must think about them in both space and time. Phylogenetic trees display the historical relationships between species. These trees allow ecologists and evolutionary biologists to explore the origins and fates of traits. Consider the following example. The white-haired goldenrod (*Solidago albopilosa*; cf. figure 3.1) is an extremely rare plant that occurs in about 90 tiny populations in north central Kentucky. All of these populations occur in sandstone "rockhouses" (shallow caves), a habitat where few other plants survive. Suppose you are an ecologist at a local university interested in this species. It would certainly be worth testing hypotheses about its seed ecology, water usage, and competitive abilities. Does *S. albopilosa* possess traits that allow it to survive in this special habitat where few other species can persist?

Further, suppose that your initial studies indicate that *S. albopilosa* has an extraordinarily high seed germination rate relative to several other plant species that occur near, but not in, the sandstone rockhouses. Over 95% of its seeds germinate, compared to 30–70% for the other species studied. This observation leads to questions that require knowledge of *S. albopilosa*'s evolutionary history. Is the high seed germination rate in *S. albopilosa* found in many other closely related *Solidago* species, or does this goldenrod display a new characteristic? If the former is true, *S. albopilosa* is only expressing a trait common in its relatives, and high seed germination may not represent a true adaptation. However, if no other closely related species possess high seed germination, it is likely that this character evolved in *S. albopilosa*'s ancestor in response to its rockhouse habitat.

Answering questions like these requires the use of phylogenetic techniques: interpreting similarities and differences among species across time to understand relationships. Phylogenetics grew out of the discipline of taxonomy, which is about discovering, describing, and naming new species, and constructing classifications for life on earth. Taxonomic classifications are meant to be practical, and not indicative of actual evolutionary relationships. As scholars accepted Darwin's claim that all species have common ancestors (and are therefore related in a treelike way), many researchers sought to uncover and organize these relationships. Phylogenetics involves reconstructing this "tree of life" using data from morphology, genetics, and behavior.

In 1950 the German entomologist Willi Hennig proposed cladistics, the method most phylogeneticists currently use to make phylogenetic trees. Hennig stressed that only shared,

FIGURE 3.1. *Solidago* spp.

derived traits (relatively new traits that are shared) should be used to infer relationships. In short, species that share evolutionarily "new" characteristics are likely to be closely related and are likely to share a recent common ancestor. The process is straightforward: a set of characters is examined for each species within an ingroup (the group of organisms being studied) and an outgroup (a separate group of organisms, usually selected for contrast to the traits of the ingroup). Table 3.1 shows traits of four fictional plant species for which we want to determine the phylogeny, and traits of a fictional outgroup species. We can use this information to construct a tree.

TABLE 3.1.
Trait information for four plant species and their designated outgroup.

Taxon	Flowers	Leaves	Stems	Number of chromosomes	Life cycle
Out-group	White	Entire	smooth	11	Perennial
Plant 1	Red	Entire	hairy	22	Perennial
Plant 2	Red	Entire	hairy	22	Perennial
Plant 3	White	Dissected	hairy	22	Annual
Plant 4	White	Dissected	smooth	22	Perennial

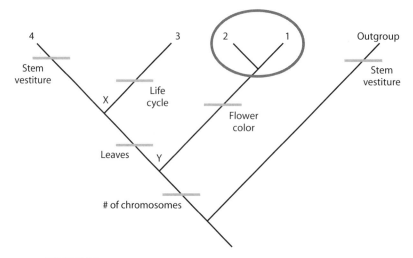

FIGURE 3.2. Phylogenetic tree constructed using the data in table 3.1.

All of the character states observed in the outgroup are likely to be relatively old, or ancestral. Ancestral traits are called plesiomorphic traits, or plesiomorphies. Character states present in the ingroup, but not in the outgroup, are likely to be relatively new, or derived. Derived traits are called apomorphic traits, or apomorphies. Figure 3.2 is a cladogram, an evolutionary or phylogenetic tree that shows groupings of species based on character states; it uses the data in table 3.1. The groupings tell us about evolutionary history. The "tips," (where you see the species numbers 1 through 4) are the present-day species we are studying, and the lines connecting them represent the branching events that we think took place in the past. Each node, or intersection, in the tree connects to an ancestor of all species on branches arising from that node. Species 3 and 4 share an ancestor at node "X." These three taxa (species 3, 4, and their ancestor X) form a clade, and all are more closely related to each other than any of them is to a nonclade member (species 1, species 2, or the outgroup). You see also that species 3 and 4 share a node ("Y") with species 1 and 2. Thus we can also call the entire ingroup a clade, a monophyletic group derived from a single ancestor.

Marks on the tree show where we think character state changes (such as the change from white to red flowers) have taken place in the past. Look carefully and you will notice that clades are formed by grouping species that have *shared, derived* characters (known as synapomorphies). For instance, the circled clade (species 1 and species 2) is identified because these two species share an evolutionarily "new" character (red flowers) that appears nowhere else. The change from white to red flowers is hypothesized to have taken place before these two species began to diverge; hence both species share this derived character. Species 3 displays an autapomorphy (in this case the fact that the plant has an annual life cycle). Autapomorphies are derived, *but not shared* by any other species, and therefore don't tell us anything about evolutionary relationships. Remember that only *shared, derived characters (synapomorphies)* tell us about these relationships. Stem vesture (presence or absence of hairs) is an example of a homoplasy, or convergence, because the outgroup and species 4 are smooth due to two separate evolutionary events, not a single event. In other words, these two species share the "smooth" character state because they converged on the same character state independently at different times in the past. A classic example

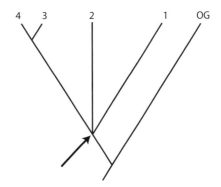

FIGURE 3.3. The arrow indicates a *polytomy*. Polytomies result when no synapomorphies (shared derived traits) can be found for a part of the tree. This tree would result if flowers of all species in the current example were the same color.

would be wings of birds, bats, and insects. It can be much more difficult to decide whether two taxa share a character because it arose in an ancestor or because of convergence. Finally, cladograms may also reflect our uncertainty. If no synapomorphies can be found in a group, part of the tree "collapses" into a polytomy in which the relationships cannot be further resolved. Figure 3.3 represents a polytomy in which we must admit that we do not know whether species 1 or species 2 is more closely related to species 3 and 4. We also do not know if species 1 and 2 are more closely related to one another than to 3 and 4.

Most characters used in cladistic analysis are discrete (e.g., white versus red versus blue flower color). Continuous characters (e.g., length of a structure) can sometimes be used in cladistics, if they can be broken up into meaningful discrete groupings (e.g., short: 1–3 cm, medium: 3–6 cm, and long: 6–9 cm).

The cladogram in figure 3.2 was constructed using the simplest solution that will explain the data; it is the most parsimonious, and therefore most likely. For a cladogram, the simplest solution is the tree with the fewest character state changes, known as "steps" (each bar on the tree in figure 3.2 is a step that marks an evolutionary change and a new character state in the descendants from that point on). Consider two alternative trees, with flower color mapped onto each tree (figure 3.4). The tree on the left is more parsimonious than the tree on the right, because it takes fewer steps to map the character state changes onto the tree. To compare two trees, we simply map each character onto each tree, and total the steps for all characters for each tree (the tree in figure 3.2 is six steps long). The tree with the fewest steps is the most parsimonious and thus most likely to represent the sequence of events that occurred and the real relationships between taxa.

Homework for this exercise takes approximately one hour.

Objectives of this Exercise

In this exercise you will

- Become familiar with basic tools of phylogenetic analysis
- Use cladistics to generate phylogenetic trees
- Use and compare morphological and molecular data in constructing phylogenies.

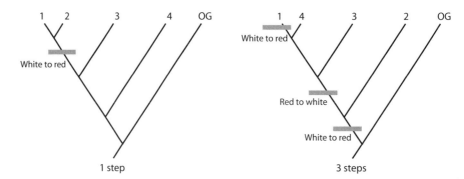

FIGURE 3.4. An example of how we choose the tree that is most likely to represent true evolutionary relationships. To compare two trees, we simply map each character state change onto each tree and sum the number of changes which would be required to result in the pattern represented on each tree. The tree with the fewest steps is the most parsimonious, and, thus, most likely to represent the true phylogeny.

Case Studies and Data

Consider the following six species. All are commonly encountered in the eastern United States. Each species is described by Whitaker and Hamilton (1998). One (the opossum) is from a very old lineage (the marsupials); it is the outgroup. The remaining five species are placental mammals, which have a well-developed internal nourishment system for developing young. Read the descriptions, looking for shared, derived characters (synapomorphies), then answer the questions at the end of the chapter.

You will use morphological, physiological, ecological and behavioral data to discover the evolutionary relationships among these mammalian species by developing a cladogram. This exercise examines living mammals, but taxonomists often work with fossils. You may wonder how we can infer so much about a fossil from skull fragments. Mammalogists have found that skull morphology, and in particular tooth morphology, can be extremely useful in distinguishing species. The number and arrangement of the different kinds of teeth are discrete characters. The mammal skull in figure 3.5(a) has three upper incisors and three lower incisors (which are difficult to see in this side view), one upper canine, one lower canine, four upper premolars, four lower premolars, two upper molars, and three lower molars. This equals 21 teeth on each side of the skull, and a total of 42 teeth in the entire skull. The dental formula tells us the number of upper and lower teeth of each type of tooth for one side. For this mammal the dental formula would be 3/3 1/1 4/4 2/3 (42). Dental formulae provide extremely useful characters for mammalogists interested in constructing cladograms.

Virginia Opossum Didelphis virginiana (Figure 3.6)

Identifying characteristics: The opossum is a cat-sized marsupial with a pointed muzzle, naked ears, and a long, prehensile tail. Long white hairs combine with black underfur.

Dental formula: 5/4 1/1 3/3 4/4 (50).

Average measurements:

total length: 762 mm
hind foot: 61 mm

Molars   Premolars   Canines   Incisors

FIGURE 3.5. (a) Side view of a typical mammal skull, with major tooth types noted. (b) This mammal jaw exhibits a diastema, a prominent gap between teeth.

tail: 306 mm
number of chromosomes: 22

Ecology and reproduction: Opossums are solitary, omnivorous, and primarily nocturnal. They occupy dens built by a variety of other mammals. Six to nine offspring are born after a gestation period of only 12–13 days in a duplex uterus. This species has a rudimentary placenta; the young nurse from nipples in an external pouch (marsupium) for up to two months.

Eastern Cottontail Rabbit Sylvilagus floridanus *(Figure 3.7)*

Identifying characteristics: This is the well-known rabbit of the eastern United States. Upper body parts are reddish brown, while the underparts (including the familiar tail) are white.

FIGURE 3.6. Virginia opossum (*Didelphis virginiana*).

FIGURE 3.7. Eastern cottontail rabbit (*Sylvilagus floridanus*).

Dental formula: 2/1 0/0 3/2 3/3 (28). These include large, ever-growing incisors, and a diastema (gap between the incisors and remaining teeth—figure 3.5(b)).

Average measurements:

total length: 414 mm
hind foot: 89 mm
tail: 49 mm
number of chromosomes: 42

Ecology and reproduction: The eastern cottontail is solitary, herbivorous, and primarily crepuscular and nocturnal. During the day cottontails sleep in a variety of shelters. Three to six offspring are born following a gestation period of 28–29 days in a duplex uterus. There is a well-developed placenta.

Norway Rat Rattus norvegicus *(Figure 3.8)*

Identifying characteristics: A brown, coarse-furred rat with a naked, scaly tail. The Norway rat is commonly found in and around human dwellings.

Dental formula: 1/1 0/0 0/0 3/3 (16). These include large, ever-growing incisors, and a diastema (gap between the incisors and remaining teeth).

Average measurements:

total length: 399 mm
hind foot: 41 mm

FIGURE 3.8. Norway rat (*Rattus norvegicus*).

tail: 187 mm
number of chromosomes: 42

Ecology and reproduction: Introduced into the United States in about 1775 from Eurasia, this pest is found in every state, and is common in almost every form of human habitation. Norway rats are solitary, nocturnal, and omnivorous. They build nests in a variety of places. Six to nine offspring are born after a gestation period of 21–23 days in a duplex uterus; they have well-developed placentas.

Domestic Cat Felis catus *(Figure 3.9)*

Identifying characteristics: The cat was domesticated from the wild cat, *Felis silvestris*, as early as 7000 years ago. This familiar human companion is about 75 cm in total length, with extremely variable hair color and texture.

Dental formula: 3/3 1/1 3/2 1/1 (30). These include large, carnassial molars, specialized for shearing, and elongated canine teeth.

Average measurements:

total length: 460–522 mm
tail: 300 mm
number of chromosomes: 38

FIGURE 3.9. Domestic cat (*Felis catus*).

Ecology and reproduction: Domestic cats are almost always found in or near human habitation. This solitary and carnivorous species is active at various hours. Typically, four kittens are born after a gestation period of 62–65 days in a bicornuate uterus. There is a well-developed placenta.

Coyote Canis latrans *(Figure 3.10)*

Identifying characteristics: Similar to an adult, large-breed domestic dog, but with longer fur and a shorter tail. The tail has a black mark above and toward the base. Coat color ranges from reddish tan and brown to black, white, and gray.

Dental formula: 3/3 1/1 4/4 2/3 (42). These include large, carnassial molars for shearing, and elongated canine teeth.

Average measurements:

total length: 1075–1200 mm
hind foot: 175–220 mm
tail: 300–390 mm
number of chromosomes: 38

Ecology and reproduction: Coyotes spend the daylight hours in wooded areas; they prepare dens to raise young. This species is largely crepuscular (active at dawn and dusk), omnivorous, and social. Typically, six pups are born after a gestation period of 58–63 days in a bicornuate uterus. *Canis latrans* has a well-developed placenta.

FIGURE 3.10. Coyote (*Canis latrans*).

Black Bear Ursus americanus *(Figure 3.11)*

Identifying characteristics: Black bears actually show a wide color range, including light cinnamon, dark brown, and black. Males can weigh up to 220 kg in the wild. Females can reach 120 kg. They have rounded ears and a rather short tail.

Dental formula: 3/3 1/1 4/4 2/3 (42). These include elongated canine teeth. The carnassial molars are largely modified for grinding, not shearing.

Average measurements (females):

total length: 1200–1500 mm
hind foot: 190–240 mm

FIGURE 3.11. Black bear (*Ursus americanus*).

tail: 80–115 mm
number of chromosomes: 74

Ecology and reproduction: Black bears inhabit densely wooded areas, using hollow trees for shelter. They are solitary and omnivorous. Active times vary seasonally and they may "den up" to hibernate during the winter in northern regions. Typically, two off-spring are born after a 70-day gestation in a bicornuate uterus. *Ursus americanus* has a well-developed placenta.

Questions to Work on Individually Outside of Class

1. Construct a simple table of characters for these mammals (similar to table 3.1), and put a star next to the synapomorphies (shared, derived characters). Not all characters are useful: some are autapomorphies (no two species share them), and some, such as body size, vary too widely to be of use. Use this information to construct your cladogram of the evolutionary relationships between these species.

2. Construct a cladogram that involves the fewest evolutionary changes in the characters you have identified. Every possible tree represents a hypothesis; present the most parsimonious tree. Ignore autapomorphies and concentrate on the synapomorphies. Remember, *synapomorphies will be the most informative characters*. Keep in mind that not all parts of the tree may be resolved (there may be polytomies). Map the characters you used onto your tree.

3. Calculate the length (number of steps) of your tree.

4. What kinds of characters did you find useful? Were discrete characters generally more informative than continuous ones?

Small-Group / In-Class Exercise

Interpreting Sequence Data

When you come to class you will be asked to generate a cladogram from DNA sequence data rather than morphological characters. This approach is becoming widely used, as DNA sequence data become available for more species.

In the first part of this exercise you performed a cladistic analysis on a diverse data set. Many familiar organisms (mammals, reptiles, insects) possess endo- or exoskeletons, providing a wealth of discrete characters. However, many organisms, such as fungi and bacteria, lack these characters; for them, convenient cladistic information can be hard to come by. By the mid-1980s, DNA sequence data had become relatively cheap and easy to obtain, and many phylogeneticists began using this nearly-endless source of information for cladistic analysis. In some ways molecular data are more straightforward than phenotypic characters. Each base on a strand is a character, and only four character states are possible (A, T, G, or C). With an outgroup to determine which character states are ancestral, identifying synapomorphies is simple. Unfortunately, you don't get to look at cool skulls.

Like most of your friends, you have applied for a variety of summer research internships. Also, like most of your friends, you exaggerated some of your experience. Having completed an exercise on cladistic analysis in your ecology course, you listed "cladistic analysis" as a skill on your resume. This landed you a job with the Department of Conservation Wetlands Ecology Team (DOC-WET) and your supervisor just handed you the first assignment for the summer.

Your supervisor wants you to resolve questions about the evolutionary relationships among several carnivorous plant species. She gave you a preliminary report and a box with tissue samples from six species. In the report you discover that the following six species are related to *Dionaea muscipula*, the familiar Venus fly trap (figure 3.12). The five ingroup species (*Nepenthes alata*, *Drosera gigantea*, *Drosera stolonifera*, *Drosophyllum lusitanicum*, and *Dionaea muscipula*) are all closely related carnivorous plant species. The report assures you that you can use *Plumbago capensis* as the outgroup. The genus *Nepenthes* has historically been placed in the family Nepentheaceae, while *Drosera*, *Drosophyllum*, and *Dionaea* have been placed in a single family, the Droseraceae. It has been suggested (Chrtek et al., 1989) that *Drosophyllum* may deserve to be placed in a separate family (the Drosophyllaceae), based on many autapomorphic characters. This leads to several questions outlined in the report:

1. Do *Drosera*, *Dionaea*, and *Drosophyllum* form a clade?
2. Do *Drosera* and *Dionaea* form a clade? That is, are both more closely related to each other than either is to *Drosophyllum*?
3. *Drosera gigantea* and *Drosera stolonifera* are currently placed in the same genus. Do they form a clade?

At this point you are torn between admitting that you have never done DNA sequencing and calling in sick for the rest of the summer. Luckily you notice a reference in the preliminary report and look up the paper by Williams et al. (1994). Amazingly, this paper already has DNA sequence data for the six species from the chloroplast gene *rbcL*, which codes for part of the photosynthetic enzyme RuBisCO. The entire gene is over 1300 base pairs, but you at least have the data for a 350-base-pair piece (figure 3.13). Now at least you have a chance to complete your assignment.

First, consider these data just as you did the morphological information in the first exercise. Ignore autapomorphies and characters that don't vary. Look for shared, derived characters, and build a table of these as you did in the homework assignment. Next, use these characters to build the most parsimonious cladogram for these species. Map all of the character changes onto the cladogram. Finally, consider the three questions noted above in the context of the cladogram you constructed. Turn in your final report, including the

FIGURE 3.12. Venus fly trap (*Dionaea muscipula*).

	1	10	20	30	40
Plumbago capensis	C C T C A A C C T G G A G G T C C A C C A G A A G A A G C A G G G G C C G C G G
Drosera stolonifera	C C G C A A C C A G G A G T T C C A C C A G A A G A A G C A G G A G C A G C G G
Dionaea muscipula	C C G C A A C C T G G A G T T C C G C C A G A A G A G G C A G G A G C A G C G G
Nepenthes alata	C C T C A A C C T G G A G T T C C G C C G G A A G A A G C A G G G G C C G C G G
Drosera gigantea	C C G C A A C C A G G A G T T C C A C C A G A A G A A G C A G G A A C A G C G G
Drosophyllum lusitanicum	C C T C A A C C T G G A G T T C C A C C A G A A G A A G C A G G A G C C G C G G

	41	50	60	70	80
Plumbago capensis	T A G C T G C C G A A T C T T C T A C T G G T A C A T G G A C A C C T G T G T G
Drosera stolonifera	T A G C T G C C G A A T C T T C T A C T G G T A C A T G G A C A A C T G T G T G
Dionaea muscipula	T A G C T G C C G A A T C T T C T A C T G G T A C A T G G A C A A C T G T G T G
Nepenthes alata	T A G C T G C C G A A T C T T C T A C T G G T A C A T G G A C A A C T G T G T G
Drosera gigantea	T A G A T G C C G A A T C T T C T A C T G G T A C A T G G A C A A C T G T G T G
Drosophyllum lusitanicum	T A G C T G C C G A A T C T T C T A C T G G T A C A T G G A C A A C T G T G T G

	81	90	100	110	120
Plumbago capensis	G A C C G A T G G A C T T A C A T G G T T C G A T C G T T A C A A A G G A C G A
Drosera stolonifera	G A C T G A T G G A C T T A C C A G C C T T G A T C G T T A C A A G G G G C G A
Dionaea muscipula	G A C C G A T G G A C T T A C C A A C C T T G A T C G T T A C A A A G G G C G A
Nepenthes alata	G A C C G A T G G A C T T A C C A G C C T T G A T C G T T A C A A G G A C G A
Drosera gigantea	G A C T G A T G G A C T T A C C A G C C T T G A T C G T T A C A A G G G G C G A
Drosophyllum lusitanicum	G A C A G A T G G A C T T A C C A G C C T T G A T C G T T A C A A A G G A C G A

	121	130	140	150	160
Plumbago capensis	T C G T A C C A C A T C G A C G C T G T T C C T G G G G G C G A A A A T C C A T
Drosera stolonifera	T G C T A C C A C A T C G A G C C T G T T G C T G G C G A A G A C A A T C A A T
Dionaea muscipula	T G C T A C C A C A T C G A G C C T G T T G C T G G A G A A G A A A A T C A A T
Nepenthes alata	T G C T A C C A C A T C G A G C C T G T T G C T G G A G A A G A A A A T C A A T
Drosera gigantea	T G C T A C C A C A T C G A G C C T G T T G C T G G C G A A G A C A A T C A A T
Drosophyllum lusitanicum	T G C T A C C A C A T C G A G C C T G T T G C T G G A G A A G A A A A T C A A T

FIGURE 3.13. 350 bases of DNA sequence for the photosynthetic enzyme RuBisCO in six species of carnivorous plants.

```
        161           170           180           190           200
Plumbago capensis       T T A T T G C T T A T G T A G C T T A C C C A T T A G A C C T T T T T G A A G A
Drosera stolonifera     A T A T T G T T T A T G T A G C T T A C C C A T T A G A C C T T T T T G A A G A
Dionaea muscipula       A T A T T G C T T A T G T A G C T T A C C C A T T A G A C C T T T T T G A A G A
Nepenthes alata         A T A T T G C T T A T G T A G C T T A C C C A T T A G A C C T T T T T G A A G A
Drosera gigantea        A T A T T G T T T A T G T A G C T T A C C C A T T A G A C C T T T T T G A A G A
Drosophyllum lusitanicum T T A T T G C T T A T G T A G C T T A C C C A T T A G A C C T T T T T G A A G A

        201           210           220           230           240
Plumbago capensis       A G G T T C T G T T G C T A A C A T G T T T T C T T C C A T T G T G G G T A A T
Drosera stolonifera     A G G T T C T G T T A C T A A C A T G T T T A C T T C T A T T G T T G G G A A T
Dionaea muscipula       A G G T T C T G T T A C T A A C C T C T T T A C T T C C A T T G T G G G T A A T
Nepenthes alata         A G G T T C T G T T A C T A A C A T G T T T A C T T C C A T T G T G G G T A A T
Drosera gigantea        A G G T T C T G T T A C T A A C A T G T T T A C T T C C A T T G T T G G T A A T
Drosophyllum lusitanicum A G G T T C T G T T A C T A A C A T G T T T A C T T C C A T T G T G G G T A A T

        241           250           260           270           280
Plumbago capensis       G T A T T T G G G T T C A A A G C C C T G C G T G C T C T A C G T T T G G A G G
Drosera stolonifera     G T A T T T G G G T T C A A G G C T C T G C G T G C T C T A C G T T T G G A G G
Dionaea muscipula       G T A T T T G G G T T C A A A G C T C T G C G T G C T C T A C G T T T G G A G G
Nepenthes alata         G T A T T T G G G T T C A A A G C C C T G C G T G C T C T A C G C T T G G A G G
Drosera gigantea        G T A T T T G G G T T C A A G G C T C T G C G T G C T C T A C G T T T G G A G G
Drosophyllum lusitanicum G T A T T T G G A T T C A A A G C C C T G C G T G C T C T A C G T T T G G A A G

        281           290           300           310           320
Plumbago capensis       A T T T G C G A A T C C C T C C T G C T T A T A C G A A A A C T T T C C A A G G
Drosera stolonifera     A T T T G C G A A T C C C T C C T G C T T A T T C A A A A A C T T T C C A A G G
Dionaea muscipula       A T T T G C G A A T C C C C C C G G C T T A T T C A A A A A C T T T C C A A G G
Nepenthes alata         A T T T G C G A A T C C C T C C T G C T T A T T C G A A A A C T T T C C A A G G
Drosera gigantea        A T T T G C G A A T C C C T C C T G C T T A T T C A A A A A C T T T C C A A G G
Drosophyllum lusitanicum A T T T G C G A A T C C C C C C T G C T T A T A C A A A A A C T T T C C A G G G

        321           330           340           350
Plumbago capensis       C C C G C C T C A T G G T A T C C A A G T T G A G A G A G A
Drosera stolonifera     A C C A C C T C A C G G G A T A C A A G T C G A G A G A G A
Dionaea muscipula       C C C A C C T C A C G G T A T C C A A G T T G A G A G A G A
Nepenthes alata         A C C G C C T C A T G G T A T C C A A G T T G A G A G A G A
Drosera gigantea        G C C A C C T C A C G G T A T A C A A G T C G A G A G A G A
Drosophyllum lusitanicum A C C G C C T C A C G G T A T C C A A G T T G A G A G A G A
```

FIGURE 3.13. *continued*

cladogram, and your conclusions about the three questions. If you can wow your boss with your analytical skills and eloquent prose, perhaps she will forgive, or never discover, your earlier dishonesty and give you a chance to complete the internship and learn those lab skills.

References

Chrtek, J., Z. Slavikova, and M. Studnicka. 1989. Beitrag zur Leitbundelanordnung in den Kronl-blatten von ausgewhalten Arten der fleischfresenden Pflanzen. *Preslia* 61: 107–124.

Whitaker, J. O., and W. J. Hamilton. 1998. *Mammals of the Eastern United States.* Ithaca, N.Y.: Cornell University Press.

Wiley, E. O. 1981. *Phylogenetics: The Theory and Practice of Phylogenetic Systematics.* New York: John Wiley and Sons.

Wiley, E. O., D. Sigel-Causey, D. R. Brooks, and V. A. Funk. 1991. *The Complete Cladist: A Primer of Phylogenetic Procedures.* Lawrence, Kans.: University of Kansas Museum of Natural History Press.

Williams, S. E., V. A. Albert, and M. Chase. 1994. Relationships of Droseraceae: A cladistic analysis of *rbcL* sequence and morphological data. *American Journal of Botany* 81: 1027–1037.

4   Life History Tradeoffs in Avian Clutch Size: Interpreting Life History Data and Evaluating Alternative Hypotheses

Jon Hess

Introduction and Background

There are strong connections across ecology, life history theory, and population biology, although they may not be obvious at first. In earlier chapters, we have talked about the fact that, in any environment, individuals with different "strategies" may not all survive and reproduce equally well. That might seem like a simple statement—but it has powerful implications for competition theory, and for population ecology. You will find, not only in this chapter but throughout this book, examples in which selection acts on clutch size (or foraging efficiency, or some other characteristic). "Best" clutch sizes (or foraging strategies, etc.) are those that result in enhanced survivorship and/or reproduction. But how do we discover these strategies?

Begin at what seems like the simplest level, and ask: if differential reproduction is key to "winning" the selective race, why doesn't every individual, every time, make the maximum number of offspring it possibly can, in physiological terms? This is a question that has occupied scholars for some time. In life history terms, the answer is constrained optimization or tradeoffs. What any individual spends on mating it cannot spend on parenting; what it spends on either mating or parenting (together called reproductive effort) it cannot spend on keeping itself alive, growing, or any other maintenance activity (called collectively somatic effort). In any particular environment, there is some optimal strategy, given the tradeoffs. So, if we understand and can measure these tradeoffs, we can, for example, calculate the optimum age at first reproduction and the optimum clutch or litter size, and solve other life history problems (e.g., Stearns, 1992; Roff, 1992, 2003).

But the implications do not stop there. These tradeoffs vary not only across species (interspecific competition), but across populations within a species, and across individuals within a population (intraspecific competition). If we can understand these tradeoffs, we can answer important questions at several levels. A classic examination of this sort was done by British ornithologist David Lack in the 1950s. Why was it, he asked, that if you observed the same population of birds across time, the average clutch size changed from year to year? Same population, same environment—or was it? Similarly, although for simplicity's sake we often talk as though patterns were "species typical," Lack found variation across and within populations. So variation can include interspecific, intraspecific, and intrapopulational effects. The good news is that variation is patterned, rather than random.

At that time, many biologists held to a (somewhat illogical) view that individuals behaved in ways that benefited the species; for example, that populations "regulated" themselves.

Before you go on, think about the definition of natural selection, and see if you can understand the logical fallacy in this assumption. It is rather subtle, because many traits that help individuals survive and reproduce also have a positive impact on, for example, population growth—so they are "good" for both the individuals and the species, and it would be easy to confuse the level at which selections acts. But reproductive patterns can highlight areas of conflict. In particular, Wynne-Edwards (1962) imagined that, if an individual did not reproduce at its physiological maximum every time, it must be paying an individual cost, in order to "regulate" the population (which he saw as good at the species level). A moment's reflection will probably convince you that there are many conditions (poor food year, a young still-growing mother) under which making the maximum possible number of offspring is not reproductively profitable. This important relationship, misunderstood by Wynne-Edwards, was one Lack studied: the nonrandom nature of variation in reproductive patterns. In a very poor resource year, is it reproductively profitable for a female bird to lay a large clutch? Most or all of the chicks will die—and she has spent her reproductive investment for nothing.

Lack did a series of now classic studies on a variety of these questions. For example, he studied nesting in the Great Tit (a British relative of the American chickadee that nests in birdhouses and is easy to study). He tested the hypothesis that clutch size is a life history trait that is under selection on the individual, as individuals compete for resources. If that is true, there may be a different "best" clutch size in good years (ample food and nest sites) and bad years, and for younger versus older birds, for example.

Objectives of This Exercise

In this exercise you will:

- Examine the variation in average clutch size across bird species
- Explore why, within species, there is usually a narrow range of variation around the average clutch size.
- Propose and evaluate alternate hypotheses about life history strategies by interpreting data
- Plot the data to help you identify trends more easily

Case Studies and Data

What factors determine clutch size? Read the following descriptions of five different species, looking first at questions 1–4 for guidance. Complete table 4.1 as you read in order to answer the questions before class.

Common Flicker Colaptes auratus (Figure 4.1)

Common Flickers (*Colaptes auratus*) are ground-foraging woodpeckers distributed across all of North America, and locally abundant. They inhabit wooded regions, including the giant cactus forests of the Sonoran Desert, and Baja, California. The form seen in the northern and eastern United States is called the Yellow-Shafted Flicker; the western form is called the Red-Shafted Flicker. A "hybrid zone" runs along the rain shadow of the Rocky Mountains; this has been studied for more than a century by evolutionary biologists and ecologists

TABLE 4.1.
Life history comparison of five avian species.

Species	Age first breeding	Longevity	Diet	Mating	Parental care	Clutch size	Precocial young?	Incubation time (days)	Time to fledge	Territorial	Migration
Colaptes auratus auratus											
Pica pica											
Aix sponsa											
Perdix perdix											
Parus major											

(Moore, 1995). Flickers are monogamous and share parental care. Their eggs are lustrous white. Egg laying begins shortly after the nest is completed. One egg is laid per day, until the average clutch size of around 6–7 eggs is complete. Eggs are incubated for approximately 11 days. Young are nidicolous ("helpless"), and are fed by both parents' regurgitation. The hatchlings subsist mainly on ant larvae. Fledglings leave the nest about 24–27 days after hatching. The young begin to breed at 1 year of age. Once an adult has bred at a locale, it returns to that locale in subsequent breeding seasons. Maximum reported longevity is 9 years and 2 months. The flickers' predators are several species of hawk, which take both adults and fledglings (Moore, 1995).

Common Flickers use a range of habitats: open woodlands, savannas, and forest edges. They specialize in eating ants and beetle larvae, but switch to berries during late autumn, winter, and early spring. Flickers forage for ants and other insects by probing and hammering soils with their bills. Like all woodpeckers, they have protrusile tongues; their tongues can extend as much as 4 cm beyond the bill tip as they lap up adult and larval ants. During the breeding season, birds forage singly; one member of a pair is usually at or near the nest. In winter, they sometimes forage in small flocks feeding on fruit (Moore, 1995).

Common Flickers have a complex array of vocalizations, drumming, and tapping that are used in communication. They use a "long call" for territorial defense and courtship. Both sexes may aggressively defend both breeding territories and mates. These woodpeckers defend a small area around their nest tree, but have ranges that extend beyond that for foraging. These ranges have flexible boundaries, and ranges may overlap (Moore, 1995).

FIGURE 4.1. Yellow-Shafted Flicker *Colaptes auratus auratus*.

Most Yellow-Shafted Flicker populations (the eastern form) are migratory, funneling into the southeastern United States in late autumn, and returning to northern breeding areas in late March to early April. Southern populations are sedentary. Yellow-Shafted Flickers migrate in loose flocks that vary in size from a few individuals to more than 100 individuals (Moore, 1995).

Black-Billed Magpie Pica pica *(Figure 4.2)*

This bird drew attention from Native Americans because it often followed bison-hunting groups and lived on the remains of the hunts. Twelve subspecies of magpies exist throughout northern Europe and Asia. Their distribution covers much of Europe, from the British Isles, Scandinavia, northern Russia, and central Siberia south to the Mediterranean region and into northwestern Africa. Their ranges extend east to Iran, the Himalayas, southeastern Asia, China, Formosa, Korea, and Japan as well as North America, where the birds probably crossed via the Bering Land Bridge. In North America, Magpie distributions remain primarily in the central western and northwestern United States (Trost, 1999). Of the two North American magpies, the black-billed magpie (*Pica pica*) is the more common and widespread.

During the nonbreeding season, magpies associate with human habitats: livestock feedlots, barnyards, landfills, sewage lagoons, and grain elevators. They are generally not migratory in North America. Their habitats in the breeding season consist mainly of thickets in riparian areas often associated with open meadows, grasslands, or sagebrush for foraging. Their long tails and slow flight make it advantageous for them to keep near cover to avoid raptors (Trost, 1999).

Magpies are opportunistic, omnivorous feeders, eating ground-dwelling arthropods, seeds, and carrion; they frequently eat fly maggots rather than the meat on which maggots live. Magpie individuals observe predators, and other magpies, to learn of potential food sources.

FIGURE 4.2. Black-Billed Magpie *Pica pica.*

Magpies cache their food and have well-developed olfactory abilities to recover caches, enabling them to steal each other's caches (Trost, 1999). Magpies have a symbiotic relationship with large ungulates (including domestic species): they clean ticks and other ectoparasites from them. However, their relationship with moose may not be mutualistic. In Alberta, when magpies land on moose to feed on winter ticks, they tend to cache these ticks on bare ground. Most ticks live through this caching process. If ticks are not recovered later by magpies, they can survive to lay eggs, potentially increasing future numbers of ticks. Many moose die each spring from blood loss and other tick-related problems, and magpies scavenge their carcasses.

Magpies have a harsh-sounding vocal repertoire. Territoriality mainly involves silent treetop sitting, which denotes ownership of space. Magpies are quite conspicuous, with white flanks and black rumps. Nests can be dispersed, or approach coloniality when food is available. In Europe, magpies are more evenly spaced than in the Americas, and they maintain traditional year-round territories (Trost 1999).

Their mating system is monogamous; some pair for life. The maximum life span is reported to be 15 years and 1 month. The mean life span for males is 3.5 years; for females it is two years. Most females begin mating at one year; however, males are usually older than a year when they begin mating. Their eggs are tan or olive with brown speckles. Clutch sizes typically range from one to nine eggs. Only females incubate; males feed their mates. Eggs hatch after about 18 days. These nidicolous young fledge 24 to 30 days after hatching, leaving the nest on average 27.5 days after hatching.

Competition among nestlings for food can lead to nestling mortality; in some cases competition intensifies into sibling cannibalism. A female may eat her eggs or small young if she is faced with human disturbance at the nest. However, starvation (both direct and indirect) is the most common cause of mortality in magpies. Underweight young are much more likely to disappear than those of average or greater weight. Chicks in broods in high-quality territories, where chicks do not beg too loudly for food (Trost, 1999), suffer less predation (especially by carrion crows) than others.

Magpies have a curious social behavior that is still not well understood. They conduct "funerals" around a dead magpie. The one to discover the dead magpie begins calling excitedly, attracting more magpies, which perch in trees, calling loudly. Up to 40 individuals may gather. Some magpies will walk around the corpse, pecking at its wings or tail. The event lasts for 10–15 minutes before all participants fly off (Trost, 1999).

Wood Duck Aix sponsa *(Figure 4.3)*

This bird is a common species in riparian habitats, wooded swamps, and freshwater marshes. Wood Ducks were very abundant across North America until the late nineteenth century, when populations declined due to the expansion of cities, overharvesting, deforestation, and destruction of wetlands. The Migratory Bird Treaty Act of 1918 protected Wood Ducks from legal harvest in the United States and Canada, and populations increased steadily, leveling off in 1985. In winter, Wood Ducks from the Canadian–United States border migrate south to southern states and into northern Mexico. Wood Ducks move in small flocks (6–12 individuals), flying mostly at night. Northeastern U.S. Wood Duck populations may travel more than 1000 km to main wintering areas; from the southeastern United States, they may travel less than 500 km (Hepp and Bellrose, 1995).

The extensive distribution of streams in the eastern United States creates the single most important habitat for Wood Ducks during the breeding season. It is essential that abundant

FIGURE 4.3. Wood Duck *Aix sponsa.*

plant and invertebrate food sources are close to suitable nest sites. Wood Ducks are omnivorous; they eat seeds, fruits, and aquatic and terrestrial invertebrates. Where wetland food is scarce, ducks may seek acorns in upland groves, nuts in orchards, and grains in harvested fields. Immediately before and during egg laying, females eat invertebrates rich in protein and calcium, which are essential for egg production (Hepp and Bellrose, 1995).

Wood Ducks are socially monogamous—they have only one mate at a time. However, because males do not help care for the young, they are sometimes serially monogamous (polygynous). A female's age at first reproduction varies by region. In South Carolina, 82% of females reproduce after one year of age, but in Massachusetts only 58% of females reproduce that early, probably reflecting differences in a number of costs (food availability, costs of migration, length of "good season"). The mean survival rate to 20 years of age is 0.56 for males and 0.51 for females.

Tree cavities provide the main nesting sites for Wood Ducks. The female lays 11–23 white glossy eggs. Incubation takes 25–37 days (average=28.8 days). Once the chicks in the eggs begin "piping" (pecking at the shell from inside), the female rarely leaves the nest unless disturbed. Ducklings usually leave the nest in the morning about 24 hours after hatching. When the coast is clear, the female flies from the nest cavity to the ground or water below, and begins calling softly. Ducklings immediately climb to the entrance and leap from the nest to the female. They are not assisted in any way, and have been reported to make spectacular jumps (up to 9 m to the ground) without injury. Obviously, these chicks are precocial. Associations between young and adults may last up to 8 weeks. When they are able to fly, the young disperse short distances (0.3–10.8 km); males go further than females (Hepp and Bellrose, 1995).

Eggs are eaten by rat snakes, raccoons, fox squirrels, mink, and various woodpeckers, including the Northern Flicker. Young ducklings are eaten by Great Horned Owls, mink, snapping turtles, bullfrogs, large predaceous fish, and alligators. Adults are vulnerable to owls, mink, raccoons, and alligators, as well as red and gray foxes. Females flush from the nest at any disturbance. Ducklings respond to the female alarm calls by skittering across the water to cover while the female swims in a different direction or feigns a broken wing. Both adults and young are excellent swimmers and will occasionally dive for food to depths of up to 1 meter. Young birds can sometimes escape predators by diving (Hepp and Bellrose, 1995).

Wood Ducks are not territorial, but males will defend mates when approached too closely; this results in a small moving territory. Wood Ducks' lack of territoriality is suggested to be an adaptation, because they breed in habitats with fluctuating water levels, where temporal and spatial distribution of food varies. There are no feeding territories in the fall or spring; pairs of birds feed close together without aggressive interaction (Hepp and Bellrose, 1995).

Partridge Perdix perdix *(Figure 4.4)*

Partridges are found in cool mid-latitude lowlands, and in temperate and steppe zones, penetrating into boreal zones and the Mediterranean. The Partridge is strictly a ground bird, preferring continuous grass or brush no higher than its head. It also seeks access to bare dusty ground, fallows, ploughed land, or dunes. Its distribution extends from the British Isles to Denmark, through Norway, Sweden, Finland, and the western part of the former Soviet Union, north into Russia. Italy and Portugal also have populations (Cramp et al., 1980), and it has been introduced into northern North America. The Partridge is mainly a "resident" bird, but some populations in Eastern Europe migrate. Diet consists chiefly of plant materials; but partridges occasionally eat insects (mostly females, when feeding chicks). Partridges are crepuscular, foraging soon after dawn and before sunset. Partridges live in flocks for 7–8 months of the year (July/early August until January or February); in the breeding season they form pairs to mate and raise chicks.

Flocks usually comprise 5–15 individuals, but may sometimes be as large as 20–25. Flocks remain within their home ranges that are greatly overlapping. Flocks tolerate each other, so each flock has a moving territory surrounding it. When birds are numerous and food is short, tolerance increases, allowing flocks to get closer, but flocks do not mix. Sometimes when flocks approach each other, old males perform threat displays for up to five minutes. When agitated, birds in the flock jump and flutter their wings. Subsequent attacks

FIGURE 4.4. Partridge *Perdix perdix.*

may occur spontaneously, or may be provoked by sudden movements in the other covey. The attacks lead to chases, and to gradual separation of the groups involved. Up to 50 birds may chase each other, apparently indiscriminately, over a small area. Aggression may occur not only between members of opposing flocks, but between siblings; parents may even attack offspring. Aggression within flocks has been noted only in early autumn, when flock composition is not fully stabilized (Cramp 1980).

When two flocks converge, pairs may form. There is a sudden turmoil of aggressive and sexual interaction; males fight to assert dominance over other males, and females over other females. In the end, males and females form socially monogamous pairs. These pairs commonly reunite in subsequent seasons. In Britain, eggs are laid in late April and early May. Nests are on the ground in thick vegetation. The eggs are olive-brown, smooth and glossy. Mean clutch size is 14.6 in England and 18.3 in Finland. Incubation takes from 23 to 25 days (mean 24.8). Only females incubate, but males may help when eggs are hatching. Young are precocial, and largely self-feeding; however, both parents accompany the young. It takes 15 days after hatching for young to fly, though they can flutter off the ground at 10 days. The age of first breeding is 1 year and the oldest breeders are about 5 years old (Cramp 1980).

Partridges have many different vocalizations to advertise and threaten, to call to gather, and to warn of ground or aerial predators. Predation, particularly by red foxes, cats, and stoats, is one of the most important causes of prehatchling mortality. Chick survival to 6 weeks is highly variable and heavily dependent on food availability (Cramp 1980).

Great Tit Parus major *(Figure 4.5)*

The Great Tit, *Parus major*, is one of the most-studied small birds in the world; its wide distribution, abundance, and readiness to nest in a box make it a convenient study species

FIGURE 4.5. Great Tit *Parus major*.

(McCleery and Perrins, 1988). One simply needs to inspect nest boxes to determine demographic information: the number of pairs, clutch size, and the number of young raised. The research suitability of the tit was first appreciated in Holland in 1912, and studies of the Great Tit have continued for almost a century now. Its diet consists mainly of insects, especially in the summer; but it forages for certain seeds in winter. These small birds are known for their relatively high metabolic rate; Great Tits spend about 75% of their day feeding (Barnes, 1975).

The distribution of *Parus major* extends from Europe far into the east, through Iran and into Mongolia. There are also scattered populations in northwestern Africa. The bird is common, diurnal, and is easily trapped and marked in the winter. In the Oxford, England, district, it is a resident all year. Adults are nonmigratory. In September the birds exhibit some territorial behavior; when it gets colder they tend to flock. Autumnal displays and fighting among Great Tits have been attributed to the advantage to young birds of establishing territories before winter. By the end of September, flocks have become stable and their ranges more clearly defined; their density is about one pair per hectare (McCleery and Perrins, 1988).

Pair formation starts at the end of January, and pairings are socially monogamous. The first clutch is usually started in April (Dhondt et al., 1996). Clutch size varies from 8 to 13 eggs (mean 9). Incubation time is 12–15 days. Only the female incubates the eggs; if the male enters the nest hole, it is usually to take food to his mate. A female will raise her wings, expand her tail, and give a most impressive hiss when disturbed while brooding her eggs (Barnes, 1975). Young are nidifugous, not nidicolous—they leave the nest. However, they are altricial, not precocial, and the female continues to brood them for a day or two after hatching because they are featherless (Barnes, 1975). Spiders and their eggs are regularly given to the nestlings; grit and crushed snails are also regular items in small quantities, presumably to aid in digestion (Barnes, 1975). If a pair's clutch fails, they will usually renest together; if the first brood is successfully fledged, some pairs may start a second brood in the same season (Dhondt et al., 1996). Second broods within a season are rare, although lost clutches are often replaced (McCleery and Perrins, 1988).

The young fledge about 6 weeks after hatching (Dhondt et al., 1996). For two to four weeks after leaving the nest, the young move about in family parties (Barnes, 1975). In July or August the young birds typically join mixed-species flocks. In late summer and early autumn, several species of warblers (and sometimes Spotted Flycatchers and Pied Flycatchers) may temporarily join tit flocks (Barnes, 1975). Only a small proportion of the fledged young reach breeding age; more than half of those that breed do so only once (McCleery and Perrins, 1988). Average age at first breeding is a year (Cichof and Linden, 1975), and mean life length is 7.8 years (Barnes, 1975).

Questions to Work on Individually Outside of Class

1. Complete table 4.1, which compares life-history traits in the five species that you read about above. Do you see patterns or possible correlations across species?

2. What aspects of avian life history might be relevant to understanding or explaining variation in average clutch size either across or within species? Do you think having precocial (or nidifugous) chicks, as opposed to altricial (or nidicolous) chicks, may affect incubation time or clutch size? Explain.

3. How does feeding behavior change in different bird species during different parts of the year? What would be some reasons for these changes? How might feeding behavior affect or be affected by breeding behavior?

4. If the size of a population is stable over time, what relationship do you expect between recruitment (birth or hatching) and mortality? How would immigration and emigration factor into this equation?

Small-Group/In-Class Exercise

This week you will use the time in class to answer the following questions in your group. These questions will require you to pay close attention to details, and to assess and integrate information from many different sources. Discuss each question as a group and hand in your written answers at the end of class. Natural selection predicts that the individual with the largest genetic impact on a population will be the one that leaves the most offspring that are in turn reproductively successful. It follows that a population would quickly become dominated by individuals that leave the greatest number of reproductively successful offspring. So is it surprising when populations of birds are observed to maintain a typical clutch size that is smaller than the largest clutch size in the population? In other words, the population does not seem to be getting replaced by birds producing the greatest number of eggs. Examine the experimental data below and discuss the questions that follow in order to discover the solution to this putative paradox. The table you assembled in question 1 at home will be useful in answering many of the questions below.

5. Consider the following manipulation: A Yellow-Shafted Flicker, *Colaptes auratus*, usually lays a clutch of 6 to 8 eggs; on average one egg is laid per day. The female then incubates the whole clutch until the eggs hatch. One female laid 71 eggs in 72 days when the eggs were removed by a scientist as soon as they were laid (Phillips, 1887).

How can you reconcile this observation with the fact that undisturbed flickers lay a maximum of 8 eggs?

6. Consider the following natural experiment:

The most frequent clutch size of the Partridge in England is 15 eggs. The Wood Duck in Iowa most frequently has clutches of 11–14 eggs. Examine the data on hatchling success in relation to clutch size in these two species (table 4.2).

(a) You can assume that, within species, the birds sampled were all approximately the same size. What can you conclude if these data are representative and do not change much from year to year?

(b) How would you state the hypothesis that could be tested with these data?

7. Recall the introductory discussion of costs and benefits of individuals reproducing maximally and individuals responding to ecological conditions. Wynne-Edwards (1962) attempted to explain why animals "restrain" themselves from producing more offspring than they are physiologically capable of producing in a given breeding season. He assumed that the *only* individually selective behavior was to make the maximum physiologically possible number of offspring. In our examples, Wynne-Edwards would have described laying as many eggs as physically possible as selfish behavior that will enhance the reproduction of the bird who lays those eggs. He explained less-than-maximum reproduction as "altruistic" restraint that avoided population crashes in the future. Such hypotheses are called "group

TABLE 4.2.
Partridge and Wood Duck reproductive data.

Partridge (based on data of Lack [1947])			Wood duck (based on data of Leopold [1951])		
Clutch size	Total eggs	Percent hatched	Clutch size	Total eggs	Percent hatched
7	—	—	7, 8	7	86
8	120	91			
9	351	88	9, 10	96	85
10	1,080	91			
11	1,716	91	11, 12	194	86
12	5,340	90			
13	5,681	90	13, 14	202	88
14	9,324	90			
15	10,650	91	15, 16	124	82
16	9,552	91			
17	6,052	90	17, 18	70	93
18	4,104	90			
19	1,520	91	19, 20	59	98
20	840	91			
21–29	831	92	21–25	116	26
Total	57,202	90.4	Total	868	80

selection" hypotheses because they imply that natural selection is working between groups and that characteristics that benefit the group as a whole (even if they are costly to the individual) will be favored. "For the good of the population" and "for the good of the species" explanations are used in popular literature and natural history films to explain a wide array of animal behaviors.

(a) Consider Wynne-Edwards' hypothesis. What advantage would an individual confer on other group members, if it restrained itself and laid few (or no) eggs?

(b) If a selfish individual (that lays as many eggs as possible) were to invade a group made up of restrained altruists, what would happen to the group composition in the next generation?

(c) If only one group exists, and it is composed of both selfish and altruistic individuals, could Wynne-Edwards' group selection have an impact on the composition of the group?

(d) Now consider both Wynne-Edwards' hypothesis and Lack's alternative individual-selection hypothesis. Which hypothesis would be more useful in explaining the Partridge and Wood Duck data?

8. Consider table 4.3 (these are data on fledgling survival in Great Tits [based on Lack et al., 1957])

Table 4.3.
Survival of young Great Tits (1945–1955); mean clutch size = 9.1–10.1.

Brood size	Initial number of young	Number alive 3 months later	Percentage alive 3 months later
3–7	200	13	6.5
8	216	14	6.5
9	369	27	7.3
10	400	26	6.5
11	220	6	2.7
12–13	61	1	1.6
Total	1466	87	5.9

    (a) What can you conclude about long-and short-term consequences of different brood sizes?

    (b) How would you state the hypothesis that could be tested with these data?

9. Use the details of the next set of studies on Magpies, along with what you know of the Magpies' natural history, to answer questions a through e. In one experiment, Magpie nestlings of 0 to 3 days old were added or removed from nests to create pairs of birds with various brood sizes (table 4.4). Reproductive Output (R.O.) is the average number of young surviving to the growth stage immediately before leaving the nest and was measured for these different brood sizes. This value is in relation to initial clutch size and the natural, or manipulated, brood size. The R.O. values are in bold for the brood sizes that correspond with the initial clutch sizes chosen by the parent birds.

In a follow up study ". . . clutch size variation between different birds in the same territory and the same birds in different territories indicate[d] that 81 to 86 percent of the within-years variation in Magpie clutch size is linked to differences in territories" (Högstedt 1980). In other words, different Magpie breeding pairs using the same territory in the same year had relatively low variation in their clutch sizes. In contrast, the same pairs trying out different territories in the same year had relatively high variation in their clutch sizes. Now try answering the following questions:

TABLE 4.4.
Reproductive output (R.O.) in the Magpie.

Initial clutch size	Brood size (natural and manipulated)					
	4	5	6	7	8	9
5	0.3	**0.7**	0.5	0.3	0	—
6	1.7	1.9	**2.8**	0.8	1.2	—
7	3.5	2.3	3.1	**3.6**	2.4	0
8	2.5	3.5	3.5	4.3	**4.5**	

TABLE 4.5.
Nestling weight and survival in the Great Tit.

Weight (15th day) (g)	Number weighed	Number alive 3 months later	Percentage
8.0–12.9	23	—	
13.0	10	—	0
14.0	21	—	
15.0	46	1	1.9
16.0	58	1	
17.0	173	9	5.2
18.0	347	19	5.5
19.0	323	19	5.9
20.0	150	8	
21.0	39	5	6.8
22.0	2	—	

Source: Based on Lack (1957).

(a) Consider a pair of Magpie breeders that occupy a given territory. They have been feeding in that territory and begin laying eggs. Later, just as they finished hatching 5 nestlings, a mischievous ecologist adds another 2 Magpie nestlings. Given the data above, what do you predict will happen?

(b) What do the data in table 4.4 and the follow-up experiment suggest determines the Magpie's clutch size?

(c) Which clutch size appears to yield the highest R.O. value?

(d) Why is it that all the Magpies are not laying the clutch size with the highest R.O. value?

(e) If 81–86% of the within-year variation in Magpie clutch size can be accounted for by differences in territories, what do you think causes the remaining variation?

10. Finally, let us consider the mean weight of nestlings in relation to brood-size. Examine the data in table 4.5 and figure 4.6, from Perrins (1965) and Lack (1957). What is suggested by these data as yet another factor important in determining most frequent clutch size?

To answer the following questions, draw on the wide array of data and examples you have become familiar with in this exercise.

11. In a season with superabundant food resources, when there is no limit to the amount of food one can gather, what do you think would be the effect on most frequent clutch size?

12. Consider birds living on an island that is often subject to devastating hurricanes. Would natural selection favor a reproductive strategy of rapid maturation and perhaps explosive reproduction, or longer time to maturity, and repeated reproduction? For example, would you expect to find reproduction heavily concentrated in one or two seasons, or smaller clutch sizes in any one year, but reproduction across six or more seasons?

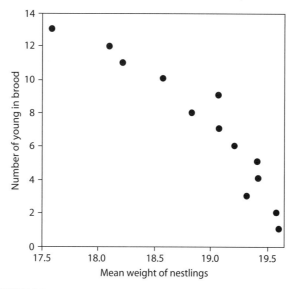

FIGURE 4.6. The relationship between brood size and nestling size.

13. What are some caveats to applying the conclusions gathered from an experiment on one species to other species (or even a single experiment within one species to all populations of a species)?

14. Integrate all the evidence you have been given and think about a wide range of bird species. What general factors determine the life history tradeoffs of large-versus-small clutch size, and few-versus-many seasons of breeding? Remember that birds must trade off their own survivorship and their reproduction. For example, fewer seasons of reproduction may result in fewer offspring over the lifetime, but if life is dangerous, early reproduction may get genes into the next generation, and ultimately result in good success in grand or great-grand offspring. On the other hand, if life is relatively safe for adults, but dangerous for offspring, adults who hedge their bets by reproducing repeatedly may prosper. Assume that resources are only mildly affected but population size fluctuates widely.

References

Barnes, J. A. G. 1975. *The Titmice of the British Isles.* London: David & Charles.

Both, C., J. M. Tinbergen, and M. E. Visser. 2000. Adaptive density dependence of avian clutch size. *Ecology* 81: 3391–3403.

Cichof, M. Z., and M. Linden. 1995. The timing of breeding and offspring size in Great Tits *Parus major. Ecology* 137: 364–370.

Cramp, S. (ed.). 1980. *Perdix perdix.* partridge. Pages 486–496 in *Handbook of the Birds of Europe, the Middle East and North Africa: The Birds of the Western Palearctic.* Vol. II. *Hawks to Bustards.* New York: Oxford University Press.

Daly, M., and M. Wilson. 1983. *Sex, Evolution, and Behavior,* 2nd ed. Boston: Willard Grant Press.

Dhondt, A. A., F. Adriaensen, and W. Plompen. 1996. Between- and within-population variation in mate fidelity in the Great Tit. Pages 235–248 in J. M. Black (ed.), *Partnerships in Birds.* Oxford: Oxford University Press.

Harrap, S., and D. Quinn. 1995. *Chickadees, Tits, Nuthatches, and Tree Creepers.* Princeton, N.J.: Princeton University Press.

Harrison, C. 1975. *A Field Guide to the Nests, Eggs, and Nestlings of British and European Birds.* New York: Quadrangle/New York Times Book Co., distributed to the trade by Harper & Row.

Hepp, G. R., and F. C. Bellrose. 1995. Wood Duck. *The Birds of North America* 169: 1–24.

Högstedt, G. 1980. Evolution of clutch size in birds: Adaptive variation in relation to territory quality. *Science* 210: 1148–1150.

Lack, D. 1947. The significance of clutch-size in the Partridge (*Perdix perdix*). *J. Animal Ecology* 16: 19–25.

Lack, D. 1954. *The Natural Regulation of Animal Numbers.* Oxford: Oxford University Press.

Lack, D. 1966. *Population Studies of Birds.* Oxford: Oxford University Press.

Lack, D., J. Gibb, and D. F. Owen. 1957. Survival in relation to brood-size in tits. *Proceedings of the Zoological Society of London* 128: 313–326.

Leopold, F. 1951. A study of nesting Wood Ducks in Iowa. *Condor* 53: 209–220.

McCleery, R. H., and C. M. Perrins. 1988. Lifetime reproductive success of the Great Tit, *Parus major.* Pages 136–153 in T. H. Clutton-Brock (ed.), *Reproductive Success.* Chicago: University of Chicago Press.

Moore, W. S. 1995. Northern flicker. *The Birds of North America* 166: 1–27.

Perrins, C. M. 1965. Population fluctuations and clutch-size in the Great Tit, *Parus major. J. Animal Ecology* 34: 601–647.

Phillips, C. L. 1887. Egg-laying extraordinary in *Colaptes auratus. Auk* 4: 346.

Trost, C. H. 1999. Black-billed Magpie. *The Birds of North America* 389: 1–27.

van Noordwijk, A. J., and J. H. van Balen. 1988. The Great Tit, *Parus major.* Pages 119–135 in T. H. Clutton-Brock (ed.), *Reproductive Success.* Chicago: University of Chicago Press.

Wynne-Edwards, V. C. 1962. *Animal Dispersion in Relation to Social Behaviour.* Edinburgh: Oliver and Boyd.

Young, H., and T. Young. 2003. A hands-on exercise to demonstrate evolution by natural selection and genetic drift. *The American Biology Teacher* 65(6): 444–448.

Mimicry: Experimental Design and Scientific Logic

James Robertson

Introduction and Background

In the evolutionary "game of life," individuals that have survived and produced the most offspring (which in turn survived and produced the most offspring, which in turn survived and produced the most offspring, etc.) are referred to as the most successful. It is their descendants that populate the earth today. To succeed at this game, organisms must acquire a variety of resources: food, water, nesting sites, and mates. Furthermore, organisms must do all this while avoiding becoming a resource themselves— e.g. through predation. Among the most fascinating antipredation traits are those associated with camouflage and mimicry. Mimicry is the close resemblance of one organism to another with the evolved function of deceiving a third organism (Lincoln et al., 1998). Mimetic adaptations can involve coloration, shape, odor, and even behavior, often in combination with one another, to create a convincing façade. Typically, mimicry involves a non-noxious species imitating a noxious one, or two noxious species converging in appearance (see below). Many species avoid detection by predators by being cryptic—blending in with their background; e.g., using camouflage. For example, a rodent's fur may, under local selection, come to match the soil or substrate color if appropriate genetic variation exists on which selection may act. For visual predators such as birds, the rodents become difficult to locate. Some of the most extreme examples of background-matching occur in the leaf and stick insects (order Phasmatoidea). These insects duplicate the appearance of leaves and sticks found in their environment (see figure 5.1). Phasmatoid insects even incorporate behavior into their façade, swaying like a leaf or stick with a gentle breeze (O'Toole, 1989). Instead of blending in with their background, other organisms use aposematic ("warning") characteristics such as bright coloration, vivid patterns, strong odors, and auditory and visual displays to bring attention to themselves. These work well if they are associated with noxious taste or other negative behavioral feedback. Aposematic characters can evolve for a variety of reasons, including sexual selection (discussed below).

In some species, aposematic characters have evolved specifically for protection by warning potential predators that the species is venomous and/or that their tissues contain poisons. Aposematic characters that truly signal toxicity or venom are called proaposematic characters. For example, a rattlesnake's (*Crotalus*) rattle warns a predator that this snake presents a true threat (venom). Similarly, the brightly colored patterns on the wings of the monarch butterfly (*Danaus*) are a real warning. As larvae, these butterflies feed on poisonous plants; the plant's toxins are incorporated into the butterfly's tissues,

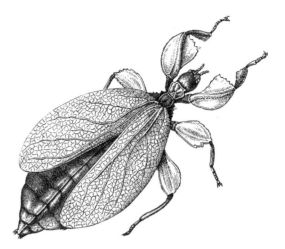

FIGURE 5.1. An insect of the order *Phasmatoidea*.

providing a chemical defense to predation. When a predator eats a poisonous *Danaus*, it vomits violently; however, the chemical seems to have no negative effect on the butterfly larvae.

The development of this defense, however, is complex. Often a predator must encounter and taste one or several individuals before associating the effects of the toxins with the pattern displayed on the butterfly's wings. A learning curve (showing improving predator response as a result of exposure) can represent graphically how a predator associates a toxic response with proaposematic traits. Predation on monarch butterflies is significantly lower than predation on closely related nonproaposematic species found in the same areas and subjected to the same predators (Ackery and Vane-Wright, 1984).

The monarch butterfly's toxic defense, using warning coloration, has been so successful that nonpoisonous butterfly species have evolved to mimic it. These nonpoisonous species have the same aposematic coloration as monarchs—but they lack the toxins. Such characters, falsely signaling venom or poison, are called pseudoaposematic characters. In other words, the species with the pseudoaposematic characters is a mimic of a species with proaposematic characters: the harmless ones mimic the dangerous. This type of mimicry (a good-tasting species mimicking a distasteful and or poisonous species) is referred to as Batesian mimicry.

The Batesian mimic relies on the "reputation" of the model species to provide itself with protection. Obviously, Batesian mimicry is frequency dependent; that is, if there are too many tasty mimics, compared to the poisonous models, predators will quickly learn that the signal (of venomousness) is a false one. As a predator encounters a Batesian mimic more frequently (as a result of an increase in the density of Batesian mimics), the predator's learning curve becomes shallower (i.e. the predator takes longer to associate the aposematic character with a threat). A shallow learning curve is disadvantageous to both the mimic and the model: more individuals of both will be harmed before the predator learns to associate the aposematic characters with a threat.

There are other cases in which two species of toxic butterflies converge in their appearance. This type of mimicry, in which both the model and the mimic are dangerous, is called Mullerian mimicry (see figure 5.2). The aposematic characters present in the mimic are

FIGURE 5.2. Monarch butterflies.

referred to as synaposematic characters. In Mullerian mimicry there usually is a difference in toxicity between the model and the mimic (the "mimic" species is often less toxic than the "model" species). However, the model and the mimic are often difficult to distinguish. For convenience, we call the species that occurs in higher densities (thus encountered more often by potential predators) the model species; the lower-density species is the mimic. In chapter 2 we introduced the concept of exaptation—that a trait evolved under natural selection for one function may later become profitable in another realm. In many sexually reproducing species, aposematic characters may be used to attract mates, thus serving a dual function. The aposematic characters may have originally evolved for the purpose of aggressive displays of intimidation both within and between species.

Objectives of This Exercise

In this exercise you will design experiments to test specific hypotheses about the function of the patterns you see. Experimental design is a whole field in itself. Bernd Heinrich's highly readable *Ravens in Winter* (1991) chronicles the extremely hard work of field biology along with the logic behind a wide array of experimental designs needed in his work with ravens.

Here we will use rather simple approaches. Nonetheless, your experiments should include: replication of experiments, randomization, and the use of experimental controls. Multiple replicates of any experiment or comparison increase our confidence that the results are the result of a real effect and not an accident. Random assignment of treatments, individuals, plots, etc. helps us avoid accidental associations or correlations that are not the result of our experimental manipulation. Finally, controls give us a population to which we can fairly compare our manipulated cases. You will be testing classic hypotheses associated with mimicry. In addition to the basic ideas and terminology used to describe and study this phenomenon, you will need some information about the study species.

Case Study and Data

One of the most impressive vertebrate examples of possible Batesian mimicry involves two salamander species. The eft, or terrestrial stage, of the red-spotted newt (*Notophthalmus viridescens*), has bright red spotting along the body, in conjunction with glands that secrete toxic chemicals. The red-backed salamander (*Plethodon cinereus*; figure 5.3) typically is dark and lacks the aposematic color pattern of *N. viridescens*. However, in some re-

FIGURE 5.3. Red-spotted newt *Notophthalmus viridescens* and red-backed salamander *Plethodon cinereus*.

gions of its distribution, *P. cinereus* has red-orange stripes on its sides and back. Although *P. cinereus* thus sometimes has aposematic coloration, it does not secrete toxins: Is it a mimic? The two species occur in sympatry (overlapping distribution) in much of the northeastern United States. Some scientists have hypothesized that *P. cinereus* is a Batesian mimic of *N. viridescens*, and that the red-orange stripping in *P. cinereus* represents a pseudoaposematic character. To test this hypothesis in the field, both color morphs of *P. cinereus*, along with *N. viridescens*, were placed in aluminum trays with high walls that prevented the salamanders from escaping. The bottoms of the trays were covered with detritus to provide the salamanders with places to hide. A third species of salamander, the dusky salamander (*Desmognathus ochrophaeus*), was placed in separate aluminum trays as an overall control, because it has neither toxins nor aposematic characters, and because it is sympatric with the other two species. In the experiment, the frequencies used for each salamander species and for each form of *P. cinereus* were approximately the same as their frequencies in nature (roughly 7 *D. orchrophaeus*, 5 *N. viridescens*, and 4 dark *P. cinereus* for every red *P. cinereus*). Individual salamanders were assigned randomly among the trays. The trays were placed in the field for two hours, exposed to naturally occurring avian predators. For each salamander species, both the initial number and the number remaining at the end of the trial were recorded (table 5.1). The experiment was (simultaneously) replicated seven times. The entire experiment was repeated later in the same location (Brodie et al., 1980).

Questions to Work on Individually Outside of Class

1. In experimental design, it is very common to include replicates of each experiment, as was done here with the seven replicates for each of the two sets (see Table 5.1). If you were designing an experiment, would you be more confident in your results with one trial or multiple trials? Explain why. How would you interpret your results if there were a great deal of variation among replicates?

TABLE 5.1.
Predation on salamanders (number offered / number remaining).

	N. viridescens	*D. ochrophaeus*	*P. cinereus* Dark form	Red form
First set of 7 (averaged)	106 / 105	145 / 77	87 / 41	12 / 6
Second set of 7 (averaged)	100 / 99	145 / 61	76 / 24	29 / 17
Average of sets 1 and 2	103 / 102	145 / 69	81.5 / 32.5	20.5 / 11.5

Source: Brodie et al., 1980.

2. The dusky salamander, *D. ochrophaeus*, was included because it was considered to be a good overall control: it has neither toxins nor aposematic coloration. What information could potentially be gained by including this species in this experiment? That is, how does the frequency with which this species suffers predation help you interpret the data on predation on other species? In general, what is the purpose of including a control in the experimental design?

3. The dark (nonaposematic) form of *P. cinereus* was also included as a control. What information would this control provide that would not be provided by the *D. ochrophaeus* control? (For example, could behavior vary between species in a way that might influence predation rates?) If behavior did vary between species, how would you use information from the dark form to determine whether behavior does influence predation rates?

4. Why did the experimental design include exposing the salamanders to the same potential predators in a second set of seven replicates?

5. In this experimental design, the frequency of individuals in the experiment differed for each species. Would the experiment have been more "fair" if there had initially been an equal number of each of the species and the two forms of *P. cinereus*? What might be the effect of having an equal frequency of mimic and model in this experiment, as opposed to frequencies that reflected empirical data?

6. Calculate the percentage of each species that survived and use those data to complete table 5.2.

TABLE 5.2.
Percent survival of salamanders.

	N. viridescens	*D. ochrophaeus*	*P. cinereus*	
			Dark form	Red form
First set of 7				
Second set of 7				
Average of sets 1 and 2				

7. To determine whether the orange spots of *N. viridescens* are indeed proaposematic characters, which two taxa would be the most appropriate to compare? Why? Does the evidence support this hypothesis? Do you need to address this before determining if the red form of *P. cinereus* is a Batesian mimic of *N. viridescens*? Why?

8. Was there an apparent difference in the survivorship of the red form between the two experimental sets? If a difference did exist, explain why.

9. Did the average percent survivorship of sets 1 and 2 (average sets 1 and 2 from table 5.2) indicate an apparent difference between *N. viridescens* and the two forms of *P. cinereus*? Why might more replicates increase our confidence that there is or is not a real difference? If a difference does exist, would this difference support the hypothesis that the red form of *P. cinereus* is a Batesian mimic of *N. viridescens*?

10. If you didn't already know that the red form of *P. cinereus* is not toxic to predators, could you make that prediction based on these experimental data? If so, which data would you use to support your prediction? If not, how would you design an experiment to test this question? If *P. cinereus* were toxic to predators, then what type of mimic would *P. cinereus* be? Why?

Small-Group/In-Class Exercise

In class today you will be integrating theory on sexual selection and mimicry with natural history data on Swallowtail butterflies. You will design subtle experimental tests of alternative hypotheses. First read the background and natural history sections below.

Background

In 1862 entomologist H. W. Bates was one of the first to describe convergent wing patterns in unrelated species of butterflies with sympatric distributions. Bates believed that there were unpalatable species that predators recognized using visual cues associated with wing patterns. If other palatable species evolved to mimic these same wing patterns, they would have reduced predation: they would be mistaken by predators for the unpalatable species. This phenomenon, later referred to as Batesian mimicry, is one of the first cited examples of Darwin's theory of evolution by natural selection (Lindstrom et al, 1997). To this day many biologists consider butterflies to be the model system for studying and understanding Batesian mimicry. However, Batesian mimicry in at least some species of butterflies has another level of complexity yet to be discussed. Within the Swallowtail butterflies (family Papilionidae) there are many examples of species that exhibit distinct differences between the sexes. A marked physical difference between males and females of the same species, sexual dimorphism, can involve a variety of traits individually or in collaboration, but is most often based on size and/or color differences between males and females.

One might predict that species relying on Batesian mimicry as a predatory defense would not exhibit sexual dimorphism. After all, how could males and females look different from one another and yet both mimic the same model species? There are, however, numerous examples of Batesian mimicry in sexually dimorphic Papilionid species; in each case mimicry is limited to females. How can this system of nonmimetic males and mimetic females be maintained, assuming that mimicry provides a selective advantage for both sexes? One possible explanation is that mimetic coloration involves a tradeoff for males. What could drive natural selection to favor nonmimetic males over mimetic males? Sexual selection has been suggested as a possible force favoring nonmimetic males (Joron and Mallet, 1998). In other words, although brightly mimetic males may have greater survival rates, other coloration may enhance reproductive success.

Sexual selection is the differential ability of individuals to acquire mates (Lincoln et al., 1998). There are two types of sexual selection: female choice and male-male competition. In sexual selection via female choice (also called *intersexual selection*), males compete indirectly with other males, as a result of females preferentially choosing certain males as mates. Females may choose males based on characters that directly indicate the male's fitness and therefore the fitness of their potential offspring (e.g., strong, active males). They may choose males based on characters that indirectly indicate male fitness because expression of the character imposes a handicap on the male; only the strongest, most energetic, or smartest males may be able to survive while dragging around the handicap (e.g., a Peacock's tail). Alternatively, females may choose males because they have characters that are attractive to other females (and their sons will be likely to express such characters and attract females as well). Hence the characters preferred by females may not be correlated with the survival component of male fitness. In the end those males with preferred traits have a higher fecundity (potential reproductive capacity) as a result of attracting more mates.

In the case of the butterflies, the females may simply prefer to mate with nonmimetic males, which are distinctly "male" in appearance. If this preference is strong enough, then even though mimetic males have reduced predation rates they might be replaced by the preferred nonmimetic males. In other words, there might be fewer nonmimetic males as a result of predation but those nonmimetic males would be responsible for most of the reproduction. Eventually, assuming a genetic basis for the coloration patterns, the nonmimetic males could replace the mimetic males in the population. (Joron and Mallet, 1998; Krebs et al., 1988).

In sexual selection via male-male competition (also called *intrasexual selection*), males compete directly with other males, in many cases forming a dominance hierarchy, for direct access to females or indirect access to females as a result of occupation of preferred territories. Male-male competition may involve aggression between males but often involves only a series of aggressive displays. Females do not choose males to mate with; instead the most aggressive or highest-ranking males prevent other males from mating. In the case of the butterflies, nonmimetic coloration or distinct male appearance may be necessary for successful aggressive displays. If nonmimetic coloration is indeed critical to aggressive displays, then even if mimetic males have reduced predation rates, they will be out-competed by the nonmimetic males (which gain access to more females). Again, if there is a genetic basis for the coloration, then nonmimetic males will eventually replace the mimetic males in the population (Joron and Mallet, 1998; Lederhouse and Scriber, 1996).

A third hypothesis, based on differential predation rates of males versus females, has been suggested to explain the combination of nonmimetic males and mimetic females. Differential predation rates between the sexes could result from the activities of one sex placing them at a higher risk to predation compared to the other sex. In the case of the butterflies, males may have a very low risk of predation when compared to females; if true, there would be some cost, and little selective advantage, to mimetic males. However, if females have a relatively high risk of predation then there may be a net selective advantage to mimicry for females. Assuming a genetic basis for mimetic coloration, selection would shape and maintain mimicry in females but not in males. As a result we would expect to see mimetic females with nonmimetic males (Ohsaki, 1995).

Natural History

There are approximately 534 species of Swallowtails (family Papilionidae) worldwide. There are 30 species found in North America (primarily concentrated in the Southern United States). Swallowtails tend to be relatively large and brightly colored. In fact, the largest butterflies in the world are in this family. The Swallowtails derive their name in part from a distinct "tail" (present in most species) which projects posteriorly from both hind wings. Swallowtail butterflies have a distinct flight pattern, flying several meters above the ground with infrequent wing beats. This results in an aerial dance of interspersed flutters and downward swoops. Although adults are capable of traveling long distances, only a few species of Swallowtails migrate. Due to their size, most species must continue to flutter their wings to support their weight while drinking nectar. When basking, these butterflies extend their wings to increase their surface area, increasing the amount of heat collected from the sun (Scott, 1986). The Pipe Vine Swallowtail (*Battus philenor*) is a distasteful model for four palatable species of Swallowtails. In the Spicebush Swallowtail (*Papilio troilus*) males and females are both Batesian mimics; there is no sexual dimorphism. In the case of both the Black Swallowtail (*Papilio polyxenes asterius*) and the Missouri Woodland Swallowtail (*Papilio joanae*) there is sexual dimorphism: only the female is a Batesian mimic of the Pipe Vine Swallowtail.

In the Tiger Swallowtail (*Papilio glaucus*) the system is even more complicated (Opler and Krizek, 1984). Tiger Swallowtail females are polymorphic; females come in two distinct forms. One is dark and mimetic, the other is light and nonmimetic. The males are always of the light nonmimetic morph. As a consequence, only females of the mimetic dark form are considered sexually dimorphic. Both males and the light morph females are yellow with three vertical stripes on each fore wing and a black border (which contains a series of yellow spots) around both pairs of wings. The dark female form (which resembles the Pipe Vine Swallowtail) is largely black dorsally and ventrally except for a distinct marginal row of spots, which are yellow above, and orange below. These spots in the dark background make this form visually very distinct to predators. The ratio of dark to light form females varies geographically, and corresponds directly to the frequency of the model species. The model species is rare in both New England and Florida, as is the dark form female. The model species is most common in the Carolinas, where Tiger Swallowtail populations consist almost exclusively of the dark female form (Opler and Krizek, 1984).

The Tiger Swallowtail is found in association with deciduous forests, although its greatest abundance occurs in riparian zones and wooded swamps. It can be found as high as 50 meters in the air moving between treetops. Males patrol particular routes, often along streams, rivers, or woodland roads, in search of females. Newly emerged males gather near muddy or sandy areas to collect mineral ions. Courtship in Tiger Swallowtails takes place in the late afternoon and involves a fluttering dance between males and females (Opler and Krizek, 1984).

Testing Alternative Hypotheses about Mimicry

For the in-class portion of this exercise, your group will design an experiment to test one of the following hypotheses about the pattern of mimetic females in association with nonmimetic males in the Tiger Swallowtails (*Papilio glaucus*).

H1. Sexual selection occurs via female choice, yielding dimorphism
H2. Sexual selection occurs via male-male competition, yielding dimorphism
H3. Differential predation rates in males versus females cause dimorphism.

You should assume that it has already been demonstrated that the dark female forms are indeed Batesian mimics of the Pipe Vine Swallowtails. You may include the other mimetic species (Spicebush Swallowtail , Black Swallowtail, and the Missouri Woodland Swallowtail) in your experiments. However, you are only required to design experiments to test one of the above three hypotheses in Tiger Swallowtails. You may wish to alter the physical appearance of the butterfly; however, if you do so, you should consider how this might influence your results. You should consider experimental design parameters such as randomization, replication and the use of controls. You can use Brodie's 1980 experiments as a general guideline for your experimental design.

You must include all experiments necessary to test your hypothesis. You should include all the potential results of your experiments, and interpret the data in each case. Be realistic in terms of the feasibility of your experiments. In other words, keep in mind that all scientific studies are limited both in time and in financial resources.

Write up your experimental plan in the form of a proposal. You will turn it in at the end of class. The proposal should include a brief abstract, the rationale behind the design, detailed materials and methods, and decision rules (i.e., how will you know if the data support the hypothesis?).

You will also have five minutes at the end of class to explain your experiment to the other groups. Think carefully about how to make your presentation clear and effective.

References

Ackery, P. R., and R. I. Vane-Wright. 1984. *Milkweed Butterflies, Their Cladistics and Biology*. Ithaca, N.Y.: Cornell University Press.

Brodie, E. D., Jr., and E. D. Brodie, Sr. 1980. Differential avoidance of mimetic salamanders by free-ranging birds. *Science* 208: 181–182.

Dunn, J. L., and E. A. T. Blom. 1991. *Field Guide to the Birds of North America*. Washington, D.C.: National Geographic Society.

Heinrich, B. 1991. *Ravens in Winter*. New York: Vintage Press.

Joron, M., and J.L.B. Mallet. 1998. Diversity in mimicry; paradox or paradigm? *TREE* 13(11): 461–465.

Krebs, R. A., and D. A. West. 1988. Female mate preference and the evolution of female-limited Batesian mimicry. *Evolution* 42(5): 1101–1104.

Lederhouse, R. C., and M. Scriber. 1996. Intrasexual selection constrains the evolution of the dorsal color pattern of male black Swallowtail Butterflies, *Papilio polyxenes*. *Evolution* 50(2): 717–722.

Lincoln, R., G. Boxshall, and P. Clark. 1988. *A Dictionary of Ecology, Evolution and Systematics*. Cambridge: Cambridge University Press.

Lindstrom, L., R. V. Alatalo, and J. Mappes. 1997. Imperfect Batesian mimicry —the effects of the frequency and the distastefulness of the model. *Proceedings of the Royal Society of London, Series B* 264: 149–153.

O'Toole, C. 1989. *The encyclopedia of insects*. New York: Facts on File.

Ohsaki, N. 1995. Preferential predation of female butterflies and the evolution of Batesian mimicry. *Nature* 378(9): 173–175.

Opler, P. A., and G. O. Krizek. 1984. *Butterflies East of the Great Plains*. Baltimore: The John Hopkins University Press.

Scott, J. A. 1986. *The Butterflies of North America: A Natural History and Field Guide*. Palo Alto, Calif.: Stanford University Press.

Section II

Demography and Population Ecology

6

Life Table Analysis

Stanton Braude

Introduction and Background

Almost every important management and conservation decision depends on understanding the demography of a species. Demographic life tables summarize the timing (and probability) of survival and reproduction; these are the basic elements of life history and natural selection. Life tables help us understand the evolution of important elements of life history, from semelparity to eusociality. Life tables are also useful in projecting growth and change in populations, so that effective, practical management practices can be developed. Almost every important question requires some information from life tables. The age-specific survivorship and fertility schedules of a life table summarize vital demographic statistics. For each age (x), the life table summarizes the probability that a newborn will reach that age (l_x) and the average number of offspring that an individual will produce at that age (m_x). Because females are typically the limiting sex, demographic life tables are usually based only on females and their daughters. Any particular combination of l_x and m_x schedules will, if unchanged, eventually lead to a particular stable age distribution with constant proportions of individuals in each age category. We note, however, that it is difficult to meet these conditions of unchanged l_x and m_x.

Once we know the survivorship and fertility schedules for a population we can plot the survivorship curve, from which we can make coarse predictions about the optimal expenditure of life effort. We can also calculate the net reproductive rate, or replacement rate (R_0), of a population, which tells us whether the population is increasing in size ($R_0>1$), decreasing ($R_0<1$), or remaining stable ($R_0=1$). R_0 is the sum of the realized fertility ($l_x m_x$) for all age classes. This is a crucial population parameter used in disciplines ranging from wildlife management to epidemiology.

$$R_0 = \sum_{x=0}^{\infty} l_x m_x. \tag{6.1}$$

Life tables then ask: given the average likelihood of surviving to each age and the average fertility at each age, how would the population grow? Consider a population of insects in which individuals live only three days. Our ages (x) could be one day, two days, and three days. Our life table would start out as table 6.1. If we know that one-third of the flies die by day 2, another third die by day 3, and none live to day 4, our life table would look like table 6.2. If we know that females lay an average of 10 eggs per day (half of which will be

TABLE 6.1.
Blank life table.

x	l_x	m_x	$l_x m_x$	E_x	V_x
1					
2					
3					

TABLE 6.2.
Life table with age-specific survivorship.

x	l_x	m_x	$l_x m_x$	E_x	V_x
1	1				
2	0.66				
3	0.33				

TABLE 6.3.
Life table with age-specific survivorship and fecundity.

x	l_x	m_x	$l_x m_x$	E_x	V_x
1	1	5			
2	0.66	5			
3	0.33	5			

TABLE 6.4.
Life table with age-specific survivorship, fecundity, and fertility.

x	l_x	m_x	$l_x m_x$	E_x	V_x
1	1	5	5		
2	0.66	5	3.3		
3	0.33	5	1.6		

Note: $R_0 = 5 + 3.3 + 1.65 = 9.95$.

daughters) regardless of age, the life table will look like table 6.3. In order to calculate the replacement rate (R_0) for the population, we first calculate $l_x m_x$ for each age category and then sum them up (table 6.4). Survivorship and fertility schedules also allow us to calculate life expectancy (E) and reproductive value (v) for each age class. Life expectancy projects the likely lifetime for an individual of a given age x in a population with specified age-specific mortality. Similarly, reproductive value projects the likely future reproduction of an individual of a given age x. These parameters are as useful for wildlife managers as they are for the life-insurance actuaries who developed these models to determine life-insurance premiums for people of different ages

$$E_x = \frac{\sum_{y=x}^{\infty} l_y}{l_x}, \tag{6.2}$$

$$v_x = \sum_{t=x}^{\infty} \frac{l_t}{l_x} m_t. \tag{6.3}$$

The life expectancy (E_x) for a fly of age x answers the question: given the prevailing age-specific mortality schedule, how much longer will a fly of age x live? In our example:

$$E_1 = \frac{1 + .66 + .33}{1} = 1.99 \text{ days},$$

$$E_2 = \frac{.66 + .33}{.66} = 1.5 \text{ days},$$

$$E_3 = \frac{.33}{.33} = 1 \text{ day}.$$

The reproductive value of a female of age x answers the question: given the prevailing age-specific fertility and mortality schedules, what is the probable number of daughters a female of age x will have in the rest of her life? In our fly example:

$$V_1 = \frac{1}{1}(5) + \frac{.66}{1}(5) + \frac{.33}{1}(5) = 9.95,$$

$$V_2 = \frac{.66}{.66}(5) + \frac{.33}{.66}(5) = 7.5,$$

$$V_3 = \frac{.33}{.33}(5) = 5.$$

Homework for this exercise takes approximately 45 minutes.

Objectives of This Exercise

In this exercise you will create a life table for a new strain of mouse, using basic descriptions of life history events. You will also use the table as a tool to project the growth of the population and determine at what point it reaches a stable age distribution. Finally, you will apply these data to the problem of controlling the growth of the population.

Case Study and Data

A group of PETA "freedom fighters" recently broke into the lab of Dr. Jacques Diani at the University of Illinois to liberate lab animals. Dr. Diani has been studying a strain of hybrid "killer" Europeanized field mice that have been spreading down the Great Plains from Canada over the past few years. These mice can move through a field of grain and, in a matter of weeks, destroy the entire crop. Now they are free in the cornfields of central Illinois. The "killer" mice live fast, love hard, and die early (there are tradeoffs in any life history strategy). They live a maximum of three months. Of those born (age class 1), 20% live to the second month of life and breed. Of these age-class-2 females, 50% (or 10% of all the newborn mice) live to the third month and breed. None of these age-class-3 mice live into the fourth month. Surviving females produce an average of 10 female offspring in each of the two breeding months. The mice released by PETA included 10 pregnant two-month-old females, 20 three-month-old pregnant females, and an assortment of males of various ages.

Questions to Work on Individually Outside of Class

1. Complete the life table (table 6.5) for the killer mice based on the description above.
2. What is the net reproductive rate (replacement rate), or R_0, of this population?
3. Is this population increasing, decreasing, or stable?

TABLE 6.5.
Killer mouse life table.

Age Class: x (months)	l_x	m_x	$(l_x)(m_x)$	V_x
1		0		
2				15
3			1	

4. Project the growth of the population based on the life table you generated and the ages of the 30 individuals that were released (table 6.6). In each subsequent month you need to keep track of who survived (and moved into the next age category) and how many were

TABLE 6.6.
Projected population growth for the released killer mice.

Age class	Population (month)											
	1	2	3	4	5	6	7	8	9	10	11	12
$N1$	0	300	50									
$N2$	10	0	60									
$N3$	20	5	0									
Total N	30	305	110									

born. We have filled in the first three months to help you see how to generate this projection. In month 1 there were 0 one-month-old mice, 10 two-month-old mice, and 20 three-month-old mice. These add up to the total of 30 mice released. The two- and three-month-old mice will each give birth to 10 daughters so there will be 300 N1 (one-month-old) mice in month 2. Because there were no N1 mice in the first month, there will be no N2 mice in the second month. And because half of the N2 mice survive to the second month, there will be 5 N3 mice in the second month.

To calculate how many of each age class will be present in the third month, first figure out which females in month 2 will be having pups. Only the five three-month-old mice will have any pups, so there will only be 50 N1 pups in month 3. Since only 20% of the N1 mice will survive, there will only be 60 N2 mice in the third month. And because there are no N2 mice in the second month, there will be no N3 mice in the third month.

Complete table 6.6 by keeping track of births and survival in these mice.

5. Can a real population continue to grow infinitely? What elements of the life table will change as the population size changes? How will they change?

6. Any population with a given l_x and m_x schedule will eventually reach a stable age distribution (i.e., proportions of each age class will be constant even if the total number in each age class changes). How many months will pass before this population reaches a stable age distribution? (Completing table 6.7 and plotting your results may help you visualize this.)

TABLE 6.7.
Age distribution across generations.

Age class	Age distribution (month)											
	1	2	3	4	5	6	7	8	9	10	11	12
$N1$	0			0.94								
$N2$	0.33	0										
$N3$	0.67		0									
Total N	1	1	1	1	1	1	1	1	1	1	1	1

7. What proportion of 1, 2, 3, and 4 month olds will there be once the population reaches a stable age distribution?

8. Describe a species for which demographic data on males would be relevant to predicting the growth of the population.

Small-Group/In-Class Exercise

Bring your completed tables 6.5, 6.6, and 6.7 to class. You will use these results to do the second phase of this assignment. Your team will work on preparing your presentation during the first half of class and you will be able to present your case at the end of class.

Life Table Analysis

It is now 3 months since the killer Europeanized Canadian mice were released into the cornfields of central Illinois. Professor Jacques Diani at Southern Illinois University has been monitoring the population with mark-recapture techniques (which you will learn about in chapter 16), and reports that the population hit about 600 animals (300 females) in the second month but is now down to about 200 animals (100 females). The governor of Illinois has approved a $100,000 emergency grant to our University's Population Biology Program to help get rid of these pests before they destroy the Illinois corn crop. Unfortunately, one of the 10 p.m. news anchors in Chicago told his audience that this is a classic example of government waste. (As sometimes happens in news reports of scientific research, he twisted Diani's findings. He told the viewers that the population is already declining and, therefore, if we leave it alone it will die out on its own.)

The governor knows better and knows that we need to act now or face a disastrous year down the road. But he is dealing with a state legislature that wants to appear fiscally responsible.

9. Your team has been granted five minutes on the Illinois State Senate floor to convince the legislature to release the funds to you now.

(a) You need to convince them that the population is going to increase in the near future.
(b) You need to explain why the recent decline was only temporary.
(c) You need to explain to them how you may be able to control (and possibly wipe out) the escaped killers by targeting your efforts on one particular age class. You are constrained by the fact that any effort to control rodents is likely to be effective with only 75% of the targeted group, population, or age class.
(d) Is there a best time to target these animals? Exactly how are you going to eradicate them?
(e) Project the effect of your program on the population and compare it to the alternatives.

You have approximately half an hour to plan your argument and prepare visual aids. A graph of the proportions of each age class over time (which you projected in table 6.6) may enable your audience to see the pattern and may help convince them. You will have five minutes to make your oral presentation to the legislature. They will then question you about your arguments.

Remember, these are not scientists, and you need to explain your reasoning fully. But they are mostly lawyers, so your argument needs to be clear, convincing, and coherent!

Reference

Wilson, E. O., and Bossert, W. H. 1971. *A Primer of Population Biology.* Sunderland, Mass.: Sinauer Associates.

7 Lotka-Volterra Competition Modeling

Stanton Braude, Tara Scherer, and Rebecca McGaha

Introduction and Background

In the previous chapter we examined age-specific fertility and mortality, using the (m_x) and (l_x) columns of a life table. We saw that we can calculate the replacement rate of a population (R_0) from the sum of the realized age-specific fertility ($l_x m_x$). In the case of the killer mice you saw that a replacement rate greater than 1 ($R_0 > 1$) leads to exponential population growth. This approach and this measure are used by scholars in many fields today (e.g., epidemiology), not just by ecologists and population biologists.

The potential for exponential population explosion was first noted by Thomas Malthus in 1798. Darwin's term "struggle for survival" grew out of his attempt to reconcile the fact that populations have the *potential* for exponential growth, yet typically remain relatively stable over time. In chapter 3 of *The Origin of Species*, Darwin notes first that he uses the term "struggle for existence" in a "large and metaphorical sense," including not just predation but competition for resources. He then noted that:

> There is no exception to the rule that every organic being naturally increases at so high a rate, that if not destroyed, the earth would soon be covered by the progeny of a single pair. . . . The elephant is reckoned to be the slowest breeder of all known animals, and I have taken some pains to estimate its probable minimum rate of natural increase: it will be under the mark to assume that it breeds when thirty years old, and goes on breeding till ninety years old, bringing forth three pairs of young in this interval; if this be so, at the end of the fifth century there would be alive fifteen million elephants, descended from the first pair.

We would be buried in elephants! Clearly, our model of population growth needs to include the impact of such forces as competition and predation, which affect R_0.

Intraspecific Competition

In 1838 Pierre-Francois Verhulst proposed the logistic growth model (later rediscovered by Raymond Pearl in 1920), which adds intraspecific competition to the exponential growth model. In the simple exponential growth model, the rate of population growth, dN/dt, continues to increase with increasing population size. The rate of growth equals the product of population size N and intrinsic growth rate r (equation 7.1).

$$dN/dt = rN \qquad (7.1)$$

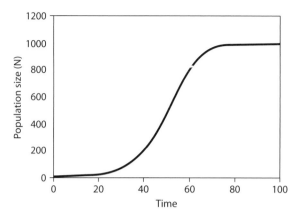

FIGURE 7.1. Example of Verhulst-Pearl population growth. Note the upper limit on population size, representative of K.

Note that r is not the same as R_0, which we derived in chapter 6. R_0, the net reproductive rate, is the sum of the age-specific fertilities, the $l_x m_x$ column, in a life table. You can think of R_0 as the average number of daughters produced by females in a generation; if $R_0 > 1$, the population is growing. The intrinsic growth rate r is analogous to an interest rate: the larger your capital, the faster your wealth grows, and the larger your population, the faster it can grow in an unlimited environment.

In the logistic model, competition also increases as a population gets larger and approaches the upper limit of population size or carrying capacity (K). Hence, the rate of growth, dN/dt, approaches zero as the population size N approaches carrying capacity. (equation 7.2).

$$dN/dt = rN((K-N)/K).\qquad(7.2)$$

Note that as the population size approaches the carrying capacity (as N approaches K), the term $(K-N)/K$ gets ever smaller, which leads to a decreasing rate of population growth. However, when population size is small in comparison to K, $(K-N)/K$ is close to 1, which leads to exponential growth. Take a moment and assume $K=1000$. Now see how $(K-N)/K$ changes if $N = 1$, 100, or 999. You can see how the shape in figure 7.1 emerges.

Interspecific competition

The *intraspecific* competition that results in a carrying capacity for a population depends in part on the resources for which conspecifics compete. Such resources define a substantial part of the **niche** of any species. *Interspecific* competition results when the niches of different species overlap, and individuals compete with individuals of another species for resources (note that both intraspecific and interspecific competition can occur simultaneously). The degree of niche overlap obviously can vary, but Gause suggested that, when overlap is significant, two species cannot coexist. His *competitive exclusion principle* specifically predicts that two species with identical niches cannot coexist in the same place at the same time; one will be driven to local extinction.

The competitive exclusion principle was tested in a classic set of experiments on flour beetles—*Tribolium confusum* and *Tribolium castaneum* (Park, 1954). The two species were grown together in a mixture of flour and yeast. One species typically ended up driving the other to extinction. When Park examined the growth conditions, he found that variations in temperature and humidity affected the outcome of the competition. *T. confusum* predominated in cool, dry conditions, while *T. castaneum* prevailed in warm, wet ones. Hence, although the two species competed for the same resources (reflected in the carrying capacity *K* of the logistic equation), their niches did not overlap with regard to optimal temperature or humidity. The carrying capacity and growth rate of each species differed under different climatic conditions. Park's results highlight an important point. As you will see shortly, there are four possible outcomes to any simple two-species competition: species A goes extinct; species B goes extinct; stable coexistence; or unstable coexistence. Ecologists puzzled for years why most experiments ended in one species going extinct while coexistence was the norm in natural situations. Part of the answer is that most competition experiments provided very simple environments. If Park had run his experiments without varying temperature and humidity, he would not have made his most important observations.

In their model, Lotka and Volterra modeled the impacts of both interspecific and intraspecific competition. In this version of the model, the impact of species 2 on species 1 is represented by the term α. The impact of species 1 on species 2 is represented by the term β. Hence, the larger the population of species 2, the greater the force of α; the larger the population of species 1, the greater the force of β The Lotka-Volterra interspecific competition model predicts that the growth of a population of species 1 in competition with a population of species 2 will be predicted by equation 7.3, and the growth of a population of species 2 in competition with a population of species 1 will be predicted by equation 7.4.

$$dN_1/dt = rN_1((K_1-N_1-\alpha N_2)/K_1), \tag{7.3}$$

$$dN_2/dt = rN_2((K_2-N_2-\beta N_1)/K_2). \tag{7.4}$$

Interspecific Competition Isoclines

We can visualize the impact of interspecific competition and predict whether a population will increase or decline for any combination of population sizes of the competing species by generating isocline plots for equations 7.3 and 7.4. The term isocline refers to a series of points (representing the population sizes of each species) along which one population is balanced between growth and competition. At other points, competition prevails and the population will decline, or growth prevails and the population grows (figure 7.2). The trick to reading these graphs easily and correctly is to remember that the size of population 1 is on the *x*-axis (so increases and decreases in population 1 move right and left). The size of population 2 is on the *y*-axis (so increases and decreases in population 2 are up and down in the two-dimensional space).

Suppose that species 1 and species 2 coexist. In this case, the population growth rate of Species 1 will be limited by both the carrying capacity due to intraspecific competition and the interspecific competition with species 2. In the Lotka-Volterra model, there are four possible outcomes that are based on the four ways the species experience the two types of competition. Take a moment to understand how these outcomes might look when we combine figures 7.2a and 7.2b.

Population 1 will drive population 2 to extinction when $\alpha < K_1/K_2$ and $\beta > K_2/K_1$
Population 2 will drive population 1 to extinction when $\alpha > K_1/K_2$ and $\beta < K_2/K_1$
The two species may coexist in equilibrium when $\alpha < K_1/K_2$ and $\beta < K_2/K_1$

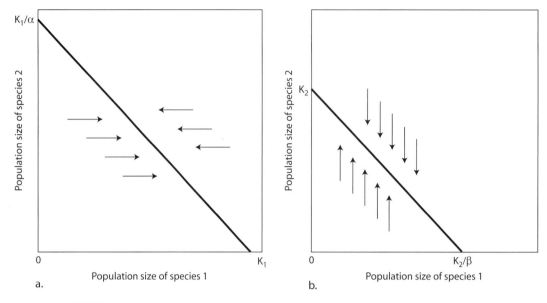

FIGURE 7.2. (a) The isocline for species 1 represents the combination of intra- and interspecific competition and the effect on the growth or decline of species 1. This graph gives no information about the impact of species 1 on species 2. K_1 = carrying capacity of the environment for species 1 alone. Arrows predict the change in population size for species 1 above or below the isocline. (b) The isocline for species 2 represents the combination of intra- and interspecific competition and the effect on the growth or decline of species 2. K_2 = carrying capacity of the environment for species 2 alone. This graph gives no information about the impact of species 2 on species 1. Arrows predict the change in population size for species 2 above or below the isocline.

Either species could drive the other to extinction when $\alpha > K_1/K_2$ and $\beta > K_2/K_1$ (the winner will be determined by the initial combined population sizes, basically who is first to get the upper hand).

That is, the results depend on whether each species is more strongly affected by competition from conspecifics or by competition from individuals of the other species. Figure 7.3 shows four plots representing the combination of isoclines for both competing species under the four conditions described above. Each isocline tells you when (at what combination of population sizes) each species is predicted to increase or decline.

Objectives of This Exercise

In the first part of the exercise, the Lotka-Volterra equation is applied to experimental data to determine population growth rates and carrying capacities. The second portion of this exercise examines the effects of interspecific competition on population growth rates.

Case Study and Data

One of the first quantitative studies of competition was conducted by G. F. Gause in microcosms with various species of *Paramecium* and rotifers. Paramecia are single-celled cili-

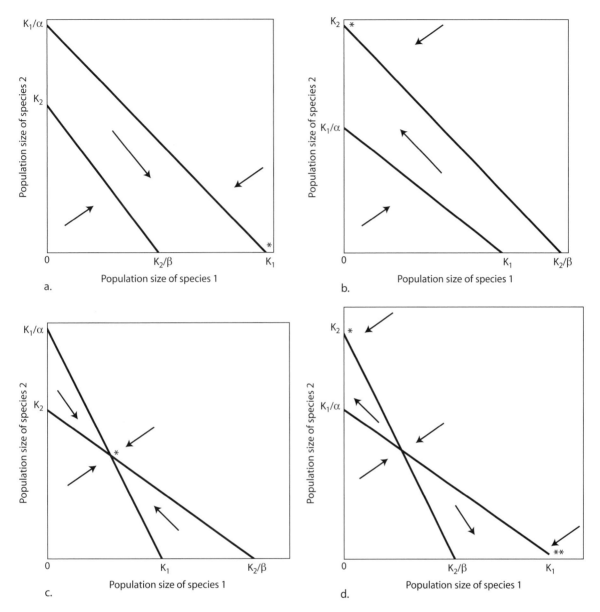

FIGURE 7.3. (a) Species 2 is driven to extinction and stable equilibrium is reached when species 1 reaches its carrying capacity at point *. (b) Species 1 is driven to extinction and stable equilibrium is reached when species 2 reaches its carrying capacity at point *. (c) Species 1 and 2 coexist in stable equilibrium at point *. Look carefully at this and figure 7.3d: They look superficially alike. Here, individuals are most affected by conspecifics, so there is a *stable equilibrium*. (d) In this, noncon-specifics have more impact than conspecifics. The system is very sensitive to the particular numbers of individuals of each species. There are three possible outcomes, but the only stable outcomes are *extinction of species 1 with species 2 at carrying capacity at point *; or extinction of species 2 with species 1 at carrying capacity at point **. The equilibrium with coexistence of both species is unstable.

FIGURE 7.4. *Paramecium aurelia* (left) is much smaller than *P. caudatum* (right).

ates commonly found in warm freshwater bodies such as ponds. They consume primarily bacteria, but also feed on yeast, algae, and other small protozoa through phagocytosis. *Paramecia* are model organisms for research, and have been the basis for extensive genetic analysis.

In one set of experiments Gause examined two common species—*Paramecium aurelia* and *Paramecium caudatum*. These two species are very similar and compete directly for food. The only major difference is in size—*P. caudatum* is about four times larger than *P. aurelia*. The two species also have different intrinsic reproductive rates r (*P. caudatum*, $r = 1.0$, *P. aurelia*, $r = 1.10$) (figure 7.4). To examine the effects these two species had on

TABLE 7.1.
Population sizes of *P. aurelia* and *P. caudatum* when cultured separately.

Day	*P. aurelia* N per 1 ml culture (cultured alone)	*P. caudatum* N per 1 ml culture (cultured alone)
0 (seeded culture)	10	10
1	20	20
2	50	80
3	150	100
4	370	160
5	530	230
6	630	235
7	740	280
8	800	240
9	900	290
10	910	320
11	840	250
12	880	280
13	980	260
14	920	210
15	800	220
16	940	250

TABLE 7.2.
Population sizes of *P. aurelia* and *P. caudatum* when cultured together.

Day	*P. aurelia* N per 1 ml culture (cultured together)	*P. caudatum* N per 1 ml culture (cultured together)
0 (seeded culture)	10	10
1	30	20
2	70	40
3	140	70
4	280	100
5	330	80
6	340	70
7	520	80
8	500	80
9	540	70
10	570	50
11	600	60
12	605	30
13	640	30
14	670	40
15	720	15
16	600	5

each other, Gause devised an experiment in which the species were grown both separately and then together in an "oaten medium" consisting primarily of diluted liquid oatmeal. This medium was inoculated with bacteria, *Bacillus subtilis*, which both species of *Paramecium* readily consumed, and allowed to incubate for a week to encourage growth of that prey population. Then, the paramecia were introduced and the tubes were closed and kept moist. Population data were collected daily and are included in tables 7.1 and 7.2.

Questions to Work on Individually Outside of Class

Problem 1

1. Plot the *P. aurelia* and *P. caudatum* logistic growth curves on separate graphs with time on the *x*-axis and population size on the *y*-axis.

2. Examine the plots and estimate carrying capacity (K) for each species.

3. What elements of the microcosm environment might affect the value of K?

4. Using the Verhulst-Pearl equation, calculate *dN/dt* for *P. aurelia* and *P. caudatum* on day 8. Which population is increasing at a faster rate on day 8?

Problem 2

5. Plot the growth of both species together on the same graph, with time on the *x*-axis and population size of both species on the *y*-axis. This will give one population growth line for each species.

6. What appears to be happening to the *P. caudatum* population by day 16? The *P. aurelia* population?

7. Now plot the populations of *P. aurelia* and *P. caudatum* in relation to each other with *P. aurelia* on the *x*-axis and *P. caudatum* on the *y*-axis.

8. As the *P. aurelia* population grows, what happens to the *P. caudatum* population?

9. If the effect of *P. aurelia* on *P. caudatum* is $\beta = 0.4$ and the effect of *P. caudatum* on *P. aurelia* is $\alpha = 2$, sketch the two isoclines onto this graph. Describe how these isoclines help illustrate the results in table 7.2, and how they help us predict the outcome of any attempt to grow these two species in the same culture tube.

Small-Group/In-Class Exercise

Bring your graphs and written answers to questions 1–9 to class along with table 7.3. In class you will examine these data, which were collected in Gause's lab with a third species, *Paramecium bursaria.* Your job will be to interpret these data in light of Gause's competitive exclusion principle. You will also be asked to revise the Verhulst-Pearl equation to include the effects of predator populations on prey populations and prey populations on predator populations.

TABLE 7. 3.
Population growth data for *P. aurelia* and *P. bursaria* cultured together. Data are reported in numbers of individuals per 1 ml.

	Trial 1		Trial 2		Trial 3		Trial 4	
Day	*aurelia*	*bursaria*	*aurelia*	*bursaria*	*aurelia*	*bursaria*	*aurelia*	*bursaria*
0	600	20	960	50	40	20	100	500
1	960	40	900	140	160	80	120	640
2	1060	80	880	260	280	140	125	720
3	1050	120	800	340	380	260	140	760
4	940	160	740	400	290	320	240	725
5	900	220	680	540	240	420	250	680
6	740	260	600	640	245	480		
7	540	320	540	700	280	620		
8	400	400	460	840				
9	300	600	310	700				

Working in Dr. Gause's Lab (Questions 10–14)

Imagine it is 1934 and you have just obtained a summer internship in the lab of Professor G. F. Gause. *Paramecia* weren't quite what you had in mind when you volunteered to work as an ecology research assistant this summer, but jobs are hard to find during the depression and at least you can live in the dorm for free. Dr. Gause has just published research on *P. aurelia* and *P. caudatum* that supported his competitive exclusion principle: "Two species occupying the same niche cannot coexist in the same place; one will always outcompete the other and drive it to local extinction." One of your first jobs was plotting those data; you have brought your results with you today.

10. Compare your results with those of the other summer interns in your group, to be sure that you all understand how Gause's data fit the Lotka-Volterra model. Which of the combined isoclines (figures 7.3a–7.3d) do your data fit if *P. aurelia* is species 1 and *P. caudatum* is species 2?

Professor Gause has been criticized for overgeneralizing his results in proposing the competitive exclusion principle. Just because *P. aurelia* and *P. caudatum* cannot coexist in many combinations doesn't mean that two similar species can never coexist. So he has decided to demonstrate competitive exclusion with a wide array of *Paramecium* species. He has assigned you to work on *P. bursaria*. Your assignment is to use *P. bursaria* to provide more supporting data for the competitive exclusion principle. You have started by growing *P. bursaria* together with *P. aurelia*. Your first trial didn't turn out as expected, so you used larger starting populations of the two species in the second trial. Again, you didn't get the expected outcome, so you try starting with really small populations of both species. Again, the expected results are not achieved, so you try starting a culture with lots more *bursaria* than *aurelia*. The data for four trials are found in table 7.3.

11. These data are not at all what your professor expects! You have a meeting with Dr. Gause tomorrow and are terrified that your career in ecology is about to meet a very abrupt end. You are already contemplating running off to China to join Mao's long march. In a last-ditch effort you decide to plot these data. Make a graph with *P. aurelia* population size on the *x*-axis and *P. bursaria* population size on the *y*-axis. Plot each of the four trials in a different color. For each trial connect the points so that you are showing the growth path of the two populations in that trial. You should end up with four different paths starting with different combinations of the two species.

12. What are you forced to conclude from these data? Which of the Lotka-Volterra isocline models (figure 7.3) do these two species appear to fit?

13. In the course of counting the microbes in your trials, you noticed that *P. bursaria* looks a little different from *P. aurelia* (it appears slightly greenish). After trekking to the library, you learn that *P. bursaria* really is unique. It is able to form a symbiotic relationship with the algae, *Chlorella*, in its cytoplasm. When *Chlorella* undergoes photosynthesis, it produces O_2, which can be used by *P. bursaria*. Given the natural history of *P. bursaria*, is there an ecological explanation for your results? How can you explain these data to Dr. Gause in the context of his competitive exclusion principle?

14. Dr. Gause surprised you by delighting in your findings. (Unfortunately, it is 1934 and you will get no credit in the monograph that he ultimately publishes.) He gives you some freedom to design your own experiments. While in the library you have become enthralled by the tiny little barrel-shaped *Didinium* that eat *Paramecia*. Before you can start your experiments with *Didinium* you need a hypothesis, a model, to test. Although you are familiar with Lotka and Volterra's competition models, you need a model of predation. Go back to

the Verhulst-Pearl logistic equation and see if you can modify it to include the effects of predators on prey populations and prey on predator populations. Make a pair of equations. (Hint: The growth of a population, r, is the difference between the birth rate and the death rate. Predators increase the death rate of the prey population and prey increase the birth rate of the predator population.)

References

Darwin, C. 1859. *The Origin of Species,* London: John Murray.

Gause G, 1934. *The Struggle for Existence.* New York: Macmillan Press.

Lotka A, 1925. *Elements of Physical Biology.* Reprinted 1956. New York: Dover.

Park T, 1954. Experimental studies of interspecies competition. II. Temperature and humidity and competition in two species of *Tribolium. Physiological Zoology* 27: 177 – 238.

Volterra V, 1926. Fluctuations in the abundance of a species considered mathematically. *Nature* 118: 558 – 560.

Explosive Population Growth and Invasive Exotic Species

Jon Hess and James Robertson

Introduction and Background

An "exotic" species is a species that has been introduced to regions outside its historic range, often inadvertently, and usually by humans. Introduced species may disrupt existing species living in communities that have never encountered the exotic introduction. The Nature Conservncy estimates that there are over 4,000 exotic plants and 2,300 exotic animal species in the United States alone! Most alarming is the fact that we know very little about the ecological and economic impact of most of these species. Many introduced species are unable to gain a foothold for a variety of reasons; however, exotic species that are able to colonize a new region successfully can have a devastating impact on local ecology. In the United States alone, more than 97 billion dollars were spent in the twentieth century to control the 79 "worst" exotic species. These include water hyacinth, which chokes southern waterways; and African bees that replace domestic honeybees (which are essential pollinators of wild and agricultural plants). Beyond economic costs are the ecological impacts of exotics. Forty-two percent of endangered species have been affected by exotic introductions, such as the unintentionally-introduced zebra mussel.

The two most important characteristics influencing a species' success as an exotic are the capacity to disperse across long distances, and the capacity to reproduce quickly so that populations grow rapidly. Species with these characteristics are often referred to as weedy or invasive species. The term "weed" conjures an image of a plant, but there are many examples of animal counterparts. Winged female fire ants (*Solenopsis invicta*), for example, disperse and mate hundreds of feet above ground. Females store sperm that will be used to fertilize eggs laid throughout their lifetimes (McKnight, 1993). After mating, a female can travel great distances to establish an entire colony on her own, producing over 200,000 workers in 5–6 years (McKnight, 1993).

Many exotic introductions were unintended, like the rats that accompanied humans to New Zealand. However, humans have deliberately introduced exotic species for many purposes: as pets, biological control agents, ornamental species, and domestic stock. As modern modes of transportation have become more efficient, rates of exotic introduction have increased. (See further examples in chapter 14, figure 14.1.)

Sometimes we compound matters by introducing exotics to control previously introduced exotics. The rosy wolfsnail, *Euglandina rosea*, was originally introduced to Hawaii in the 1950s as a biological control agent to reduce populations of another exotic snail, the giant African snail (Stone, 1999). The rosy wolfsnail, which uses *chemotaxis* to track its prey (Stone, 1999), is unfortunately, a more generalist predator than had been expected; it eats far more species than giant African snails. It has an unfortunate preference for native Hawaiian snails (Stone, 1999) and more than 15 native snail species have been driven extinct. So far, the only means of reducing damage by the rosy wolfsnail has been the construction of special fences, lined with salt barriers, to keep the snail out of uninfested habitat (Stone, 1999).

In North America in the early 1800s, purple loosestrife (*Lythrum salicaria*) was introduced. This European plant has since established itself in wetlands in practically every state of the United States. Although there are few good data on the ecological influence of purple loosestrife, the sheer numbers and rapid proliferation of the plant have led to concern that it "crowds out" native plants such as cattails (Hager and McCoy, 1998).

Exotic species are a problem throughout the world. Australia, for example, has been waging a war against the European rabbit (*Oryctolagus cuniculus*) since the 1800s, when it was introduced by British settlers (Ross and Sanders, 1984). The infamous reproductive capabilities of the rabbit led to rapid population growth and massive defoliation of the landscape. Rabbits soon became severe competition for both native fauna and livestock (Finkel, 1999). Authorities in Australia resorted to introducing a myxomatosis virus to control rabbits in 1950. Initially, the virus was an effective rabbit killer. However, in just eight years the degree of resistance to the virus had increased from approximately 10% to 74% (Ross and Sanders, 1984). The virus was spread by mosquitoes, which will not bite dead rabbits. So, while rabbits were evolving resistance, the viral strains were evolving lower virulence because viruses that kept their hosts alive longer were able to spread more effectively. In 1995 a new virus, the European rabbit calcivirus disease (RCD), was released. So far, it has been successful at lowering rabbit numbers.

Many unintentionally introduced exotics are hitchhikers in the transportation of goods. For example, the fire ant probably arrived in the United States in ships from South America. The fire ant was initially observed in Alabama in 1918, and has since expanded its range throughout southeastern states. We do not yet know the extent of fire ant damage, but we do know that fire ants have a competitive edge over native ants, and damage certain crops (McKnight, 1993).

Homework for this exercise takes approximately 45 minutes.

Objectives of This Exercise

This exercise examines the growth of exotic species populations and compares the growth curves of different populations. The second portion of the exercise examines the impacts of exotic species on native food webs and on native populations.

Case Studies and Data

Transoceanic shipping is responsible for the introduction of many exotic aquatic species throughout the world as a result of moving ballast water from port to port. Ballast water is water pumped into or out of internal tanks on ships to achieve the desired "draft" and stability, depending on the cargo load; if the load is light, ballast water is pumped in to make

FIGURE 8.1. Zebra mussel *Dreissena polymorpha.*

the ship ride lower and stably. Small aquatic organisms, including early developmental stages of large species, can be inadvertently transferred in the ballast water to new localities in this way (McKnight, 1993). In the Great Lakes alone, over 800 million liters of ballast water are discharged every year (Ricciardi and MacIsaac, 2000). The zebra mussel (figure 8.1), *Dreissena polymorpha,* is a fingernail-sized introduced transoceanic exotic. The species was once restricted to the Ponto-Caspian Basin in the southern part of the former Soviet Union. Now it is widespread throughout Europe (Lori and Cianfanelli, 2006), Great Britain, and the Great Lakes region of North America.

Zebra mussels are dioecious, meaning the sexes are separate. They begin their lives as free-swimming larvae that develop 2–3 days after the egg is fertilized. They continue their planktonic (free-swimming in the top layers of the water) development for approximately 33 days. Afterward they become sedentary (McKnight, 1993). During the sedentary stage the larvae settle on patches of solid substrate, like rock, wood, ship hulls, fishing gear, water fowl, etc. They avoid sand, mud, silt, and clay (Ludyanskiy et al., 1993). After about 11 months in this sedentary stage, the mussels become sexually mature. The sedentary period typically peaks in August, and spawning, the reproductive phase, typically peaks in July. Females expend an extraordinary amount of energy producing around 300,000 eggs during the course of a spawning season; of these, only 1% are believed to survive to the sedentary stage (McKnight, 1993). After the reproductive phase, most of the adults die as a result of their exhaustive reproductive effort. However, a few survive to the next spawning season.

This high rate of reproduction, together with effective long-range dispersal, has led to the rapid colonization of zebra mussels throughout U.S. waterways. The zebra mussel was first introduced into the United States in 1986, as a result of ballast water discharged in the St. Lawrence Seaway in the Great Lakes region. By 1989, zebra mussels were abundant in both Lake Erie and the western part of Lake Ontario, and by 1990, they had extended their distribution into all of the upper Great Lakes as well as into the St. Lawrence River and New York State Barge Canal. In 1991, the mussels were found in the Hudson, Illinois, Mississippi, Ohio, and Tennessee rivers (McKnight, 1993).

Negative impacts of the zebra mussel include economic as well as ecological disruption. Economic costs result from fouled water pipes and restricted water flow. Power plants depend on water, and when the mussels reduce water flow (by up to 80% in some pipes) the result is costly plant shutdowns (McKnight, 1993). Ecological damages include the extinction of native bivalves, and the crippling of food webs through removal of large portions of

TABLE 8.1.
North American populations of the zebra mussel *Dreissena polymorpha*.

Lake Michigan		Point Pelee, Lake Erie		Missouri River	
Month	thousands/m²	Month	thousands/m²	Month	thousands/m²
Jan-93	4.021	Jan-91	8.762	Jan-92	0.1201
Feb-93	4.966	Feb-91	8.884	Feb-92	0.1962
Mar-93	5.277	Mar-91	9.011	Mar-92	0.3669
Apr-93	5.443	Apr-91	9.761	Apr-92	0.7953
May-93	5.743	May-91	9.344	May-92	1.6386
Jun-93	6.977	Jun-91	9.875	Jun-92	2.7950
Jul-93	6.755	Jul-91	10.045	Jul-92	2.5346
Aug-93	7.563	Aug-91	11.353	Aug-92	1.8057
Sep-93	7.859	Sep-91	10.863	Sep-92	1.0452
Oct-93	8.214	Oct-91	8.421	Oct-92	0.9254
Nov-93	6.982	Nov-91	3.521	Nov-92	0.4219
Dec-93	5.471	Dec-91	1.232	Dec-92	0.5947
Jan-94	5.683	Jan-92	0.657	Jan-93	0.5836
Feb-94	4.695	Feb-92	2.347	Feb-93	0.3427
Mar-94	4.157	Mar-92	3.337	Mar-93	0.9038
Apr-94	4.569	Apr-92	3.212	Apr-93	1.5870
May-94	5.352	May-92	4.426	May-93	3.1627
Jun-94	6.011	Jun-92	4.938	Jun-93	3.3415
Jul-94	5.958	Jul-92	4.875	Jul-93	6.3324
Aug-94	6.423	Aug-92	3.569	Aug-93	6.3712
Sep-94	5.870	Sep-92	3.397	Sep-93	4.9726
Oct-94	4.783	Oct-92	2.594	Oct-93	2.1664
Nov-94	3.986	Nov-92	2.237	Nov-93	2.1760
Dec-94	5.347	Dec-92	2.884	Dec-93	2.1193
Jan-95	5.835	Jan-93	2.974	Jan-94	1.6294
Feb-95	6.295	Feb-93	3.325	Feb-94	2.1299
Mar-95	5.955	Mar-93	3.187	Mar-94	2.2009
Apr-95	6.334	Apr-93	3.455	Apr-94	4.8727
May-95	7.673	May-93	4.008	May-94	5.8301
Jun-95	8.311	Jun-93	5.227	Jun-94	10.254

TABLE 8.1.
(*continued*)

Lake Michigan		Point Pelee, Lake Erie		Missouri River	
Month	thousands/m^2	Month	thousands/m^2	Month	thousands/m^2
Jul-95	7.836	Jul-93	4.981	Jul-94	9.2754
Aug-95	8.288	Aug-93	3.833	Aug-94	10.315
Sep-95	6.541	Sep-93	3.015	Sep-94	13.618
Oct-95	6.397	Oct-93	2.874	Oct-94	7.8232
Nov-95	5.274	Nov-93	2.933	Nov-94	5.5337
Dec-95	3.154	Dec-93	2.359	Dec-94	3.2473

primary production (Ludyanskiy et al., 1993). Zebra mussels do not harm natives only by their efficient filtering system; they also settle on the shells of native mussels. This decreases the natives' ability to burrow or maintain a normal position (Schloesser et al., 1996).

There are over 150 species of unionid bivalves native to the United States. A third of these are included on the Federal Endangered Species list. Thus, the threat of the zebra mussel is undoubtedly serious (McKnight, 1993). The zebra mussel poses a problem to all species dependent on phytoplankton and other primary productivity that the mussels use as a food source (above); they have drastically altered local food webs. Problems are expected to increase in the future. Because zebra mussels' economic and ecological impact are severe, the Canadian and U.S. governments have already spent millions of dollars to reduce their numbers (Ludyanskiy et al., 1993).

Zebra mussels have recently moved into Missouri. The Department of Conservation has urged lawmakers to take measures to halt the explosive population growth of the invasive mollusks. However, some state representatives argue that the mussels are not growing exponentially; they cite data on zebra mussel populations in Missouri in 1992 and data from Lake Erie and Lake Michigan.

Your job is to look more closely at the real data and to help our local state representative understand what is going on in the populations. The following data were gathered by a team of scientists who sampled zebra mussels at three sites: the northern banks of Lake Erie, the southern banks of Lake Michigan, and the mouth of the Missouri River (table 8.1). They used SCUBA gear to obtain the zebra mussels, and recorded only the numbers of mussels with a shell length of 15 mm or greater (this is important because zebra mussels of this size are sexually mature). Mussel density was calculated for each site every month over a three-year period. Note that each site was visited over a slightly different three-year period, but the periods do overlap with each other.

Questions to Work on Individually Outside of Class

The zebra mussel hearings are scheduled for next week; you need to deliver your interpretation of the data by then. Your report needs to be typed with all relevant figures attached (plots of the available data will be essential to visualizing the trends within and differences between populations). Your boss has outlined the questions you need to address in your report:

1. How do you explain the fluctuations in population size among the adult mussels apparent in the data? (Remember, the only mussels taken into account are those of shell length ≥15 mm.)

2. Has the zebra mussel population in the Missouri River reached the point where it is operating under the influence of carrying capacity? How would you describe the growth of this population (stable, linear, exponential, or logistic growth)? Explain the data in terms of the trends and what is known about the life history of the mussel's dispersal.

3. Does the southern Lake Michigan population appear to have reached carrying capacity? Explain the data in terms of the trends and what is known about the life history of the mussel's dispersal.

4. How could you explain the Point Pelee data in terms of carrying capacity, given that Lake Erie experienced severe storms and flooding that dumped massive amounts of silt into the lake in late 1991?

5. How would data collected over a shorter time period affect your predictions about a population's future growth? For example, one state representative cited only the 1992 data from the local population in Missouri. What predictions would those data lead to, and how would they compare to predictions based on the complete data set?

6. Is there any risk of an explosion in the Missouri population of zebra mussels? What evidence is there in the data? How could the data have been misinterpreted by state representatives who weren't smart enough to hire a student like you as their technical aide?

7. Consider the following two possible outcomes of a genetic survey of zebra mussels. Offer a realistic scenario that could have given rise to each result.

Possible Result A: All of the alleles found in the European population are also found in the Great Lakes. In other words, the European population is genetically indistinguishable from the Great Lakes population. Possible Result B: Collections of mussels from opposite ends of Lake Erie are genetically indistinguishable. How might the reproductive biology of the species contribute to such a result?

Small-Group/In-Class Exercise

Read the following section on zebra mussels and then work through the questions on explosive population growth in order to draft your position paper on the topic.

Beneficial Effects of Zebra Mussels

Zebra mussels are extremely efficient natural biofilters; they have been intentionally introduced to lakes across Europe to keep the water clear and beautiful (Ludyanskiy et al., 1993). They filter detritus, consuming particles that are 10–40 microns in size. It would take only ten days for zebra mussels to filter all of the water in Lake St. Claire (assuming that they had attained a real life density of 50,000 mussels [all sizes included] per square meter and covered 1% of the lake bottom). This incredible filtering potential can purify lakes by removing dissolved nutrients like phosphorus, calcium carbonate, and other minerals. The zebra mussels are capable of accelerating the conversion of toxic nitrogenous metabolic wastes to consumable nutrients for other benthic organisms and phytoplankton.

Removal of excessive amounts of algae by the mussels contributes to the restoration of lakes and other bodies of water. The mussels can also control eutrophication (Ludyanskiy et al., 1993), the phenomenon in which overenrichment of soluble nutrients allows explosive growth of some populations, which in turn deplete the available oxygen. This leads to suffocation of organisms at higher trophic levels. Zebra mussels are also popular in Europe as bioindicators. The mussel thrives only in clean, oxygen-saturated water. For researchers studying ecotoxicology (effects of pollution and other toxic stress on ecosystem structure and function) in Europe, changes of the zebra mussel valve movement are a simple indicator of pollution. For a more involved investigation, zebra mussel tissues are examined for trace metals, organochlorines, and radionuclides. Zebra mussels are excellent bioindicators because they take these pollutants up readily from the water and concentrate them in their tissues (Ludyanskiy, 1993).

People may complain about the sharp shells on the lake bottom and the odor of the mussels decaying on the shore, but for some this is a small price to pay for cleaner lakes and streams.

Unionid Bivalve Infestation by Zebra Mussels

Unionid bivalves are freshwater clams and mussels. There are 297 native unionid species in the United States. Unionids are a crucial component of all freshwater communities because they are natural water purifiers and integral links in freshwater food webs. Zebra mussels tend to use the hard shells of native unionids as a place to settle. This direct infestation by zebra mussels on native unionid bivalves is suspected to cause high mortality. Hundreds of zebra mussels encrust the outer shell of unionids when the zebra mussel population densities begin to exceed 5,000/m². Infestation intensities of greater than 150 zebra mussels (one year old and older) per unionid can be fatal for the North American hosts. Mean infestation intensities in European studies tend to fall below 180 zebra mussels per unionid and coexistence has been possible (Schloesser et al., 1996). A team of researchers interested in zebra mussel infestation of unionids has collected unionid specimens over a three year period in Lake Erie. The results of their findings are reported in figure 8.2. The team has also sampled a larger area that includes the Great Lakes and major connecting rivers. Their data are reported in table 8.2.

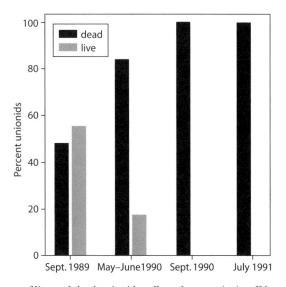

FIGURE 8.2. Percentage of live and dead unionids collected at one site in offshore waters of western Lake Erie 1989–1991. In the data presented above the maximum infestation intensities were 11,550/live unionid and 14,393/dead unionid in western Lake Erie.

Diving Ducks and Zebra Mussels

Diving ducks have received a great deal of attention in the debate over zebra mussel control. Some hope that they can serve as a native predator to keep zebra mussel populations in check. Several species of diving ducks, including the Greater and Lesser Scaup, Common Goldeneye, and Bufflehead, stop in the Great Lakes during their annual migrations. Their main source of food during these stops is the zebra mussel. While this discovery is exciting (in the sense that there is potential for natural control by the ducks) officials remain skeptical about the long-term effects of the ducks on the mussel populations. There are relatively few wintering ducks in the Great Lakes compared to the high densities on European lakes, where the ducks are effective control agents. Since 1990 researchers have been monitoring the ducks that stop at Point Pelee, Lake Erie, during fall migration. Diving ducks arrived in late October 1991 and remained in numbers approaching 1000 ducks until mid-December of 1991 (Hamilton et al., 1994). There were no ducks present before this time and few remained into the winter. The ducks migrated back in the spring. Numbers approaching 2000 were observed on April 2nd 1992. During these times ducks were found to feed almost exclusively on zebra mussels.

The researchers shot a number of ducks and found that the content of their gizzards were composed entirely of zebra mussels. The distribution of mussel sizes found in the duck gizzards is shown in figure 8.3.

The ducks receive considerable nourishment by eating these high-calorie treats. The caloric value of the mussels increases exponentially with shell length, but their shells thicken as mussels grow.

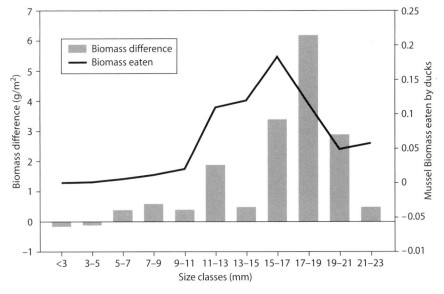

FIGURE 8.3. Difference between cage and control area biomass (cage density–control density) of zebra mussels for November north sites (bars) plotted against biomass found in duck gizzards (line). Biomass of mussels eaten by ducks was estimated for each size class by multiplying mean dry tissue mass of a mussel in that size class (as estimated from the November 1991 sample) by number of mussels from that class found in gizzards (from Hamilton *et al.*, 1994).

TABLE 8.2.
Species of Unionid mollusks infested by zebra mussels in the Great Lakes and connecting rivers 1989–1994.

*Amblema plicata plicata	LSC	WLE	PI	IMR	
Anodonta implicata					HR
*Anodontoides ferussacianus			PI		
Arcidens confragosus				IMR	
Ellipsaria lineolata				IMR	
*Elliptio complanta	LSC				HR
*Elliptio dilatata	LSC		PI	IMR	
*Fusconia flava	LSC		PI		
*Lampsilis cardium	LSC	WLE	PI	IMR	
*Lampsilis siliquoidea (& radiata)	LSC	WLE	PI	IMR	HR
Lamsilis teres				IMR	
*Lasmigona complanata complanata	LSC			IMR	
*Lasmigona costata			PI		
*Leptodea fragilis	LSC	WLE	PI	IMR	
Leptodea ochracea					HR
*Ligumia nasuta	LSC		PI		HR
*Ligumia recta	LSC		PI		
Megalonaias nervosa				IMR	
*Obliquaria reflexa	LSC			IMR	
*Pleurobema coccineum	LSC				
*Potamilus alatus	LSC	WLE	PI	IMR	
*Potamilus ohiensis		WLE		IMR	
*Ptychobranchus fasciolaris	LSC		PI		
*Pygandodon grandis	LSC	WLE	PI	IMR	
Quadrula nodulata				IMR	
*Quadrula pustulosa pustulosa		WLE	PI	IMR	
*Quadrula quadrula			PI	IMR	
*Strophitus undulatus			PI		
*Tuncilla donaciformis			PI	IMR	
*Truncilla truncata	LSC		PI	IMR	
Utterbackia imbecillus				IMR	

Note: An asterisk indicates species where mortality has been attributed to infestations. LSC = Lake St. Clair; Gillis and Mackie (1994); Hunter and Bailey (1992). WLE = western Lake Erie; Haag *et al.* (1993); Schloesser and Nalepa (1994). PI = easter Lake Erie, Preseque Isle; unpublished, DWS. IMR = Illinois and Mississipi rivers; Tucker *et al.* (1993); Tucker (1994). HR = Hudson River; Strayer *et al.* (1994); Strayer, personal communication.

To quantify the impact of ducks on the mussel populations, the researchers set a cage around an area of mussels on the lake bottom, making it impossible for the ducks to feed on the mussels in the enclosed area. This caged site was their test treatment. Beside the caged site, an open site was marked on the lake bottom. This open site was the control treatment. By comparing the number of mussels in these two treatments, the researchers were able to quantify the impact of duck predation on the mussel population. They used cages at two different sites at Point Pelee: one northern and one southern. In the fall of 1991, ducks were seen feeding and resting around the northern site, but for the most part ducks ignored the southern site. The differences between caged and uncaged areas at the northern and southern sites are shown in figures 8.4 and 8.5. All size classes were included in the density data.

Explosive Population Growth and Invasive Exotic Species

Your boss has called you in because you were the one who convinced her that the growth trajectory of the zebra mussel population at the mouth of the Missouri is in fact exponential and is likely to continue increasing. The same lawmakers who originally argued that the population in Missouri is not growing exponentially are now trying to argue that the species is not a threat, even if it is growing exponentially. They have been circulating articles about the benefits of exotic species and the potential economic and environmental problems associated with attempts to control them. They have also appealed to the hunting lobby by suggesting that zebra mussels have a positive impact on waterfowl populations, and that these waterfowl populations serve as an effective means of natural control. *Your task is to integrate your knowledge of zebra mussel population growth with new information to draft a position paper on the issue.* Answer questions 8–12 to help you understand the

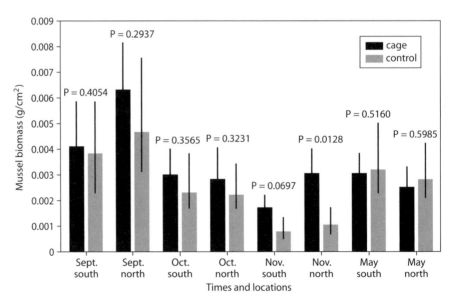

FIGURE 8.4. Zebra mussel biomass per unit area (including mussels of all size classes) for cage and control treatments from September 1991 to May 1992. Biomass is the dry weight of the flesh inside the shell. Probabilities are from one-tailed paired *t*-tests, testing the null hypothesis that control density is not less than cage density (from Hamilton *et al.* 1994).

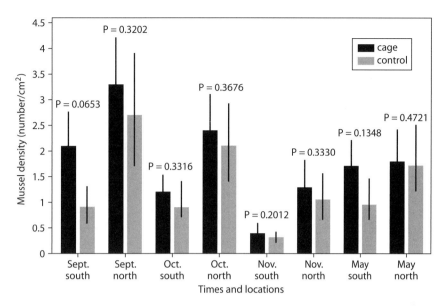

FIGURE 8.5. Zebra mussel number per unit area (including mussels of all size classes) for cage and control treatments from September 1991 to May 1992. Probabilities are from one-tailed paired *t*-tests, testing the null hypothesis that control density is not less than cage density (from Hamilton *et al.*, 1994) (from Schloesser et al., 1996).

data, then answer questions 13–15 to synthesize the various information you have at your disposal.

Duck Predation on Zebra Mussel Populations (Questions 8–11)

8. Is there is any significant difference in the zebra mussel populations between the caged areas and free areas for each of the different months and lake locations? (Note: the researchers refer to free areas as control areas on figures 8.4 and 8.5 although that label is confusing.) How does the background information you have help explain why each month did, or did not, show a significant difference in the size of the zebra mussel population? (A *p* value less than 0.05 would be considered statistically significant; see chapter 15.)

9. How could the researchers determine the zebra mussel biomass that was potentially available to diving ducks? How could they determine what fraction of that was consumed? What trends in "available" and "consumed" biomass are apparent in the data presented in figure 8.3?

10. Not all the data fit the predictions about the zebra mussel populations in Lake Erie. Which data are contrary to the predictions?

11. How does the abundance of ducks affect zebra mussel populations, and vice versa?

Zebra Mussel Infestation and Native Species (Question 12)

12. Can you conclude that "there is a correlation between infestation and high unionid mortality"? With the data available, can you conclude that "zebra mussel infestation causes high mortality in unionid populations"? What is the difference between these two statements? (see chapters 14 and 15 if these concepts are not familiar). Briefly outline an experiment that could test the hypothesis that zebra mussel infestation *causes* unionid mortality.

Synthetic Proposal for the Zebra Mussel (Questions 13–15)

13. A member of the senate decides that the introduction of zebra mussels and ducks would be ideal to restore the health of a small lake with a history of pollution from farm fertilizers and heavy metals. He suggests that zebra mussels would filter the pollutants in the pond and the presence of ducks would control the zebra mussel population. This lake is located in northern Michigan and is home to a struggling aquatic community that is sensitive to the pollutants. Show why this senator's proposition could be more trouble than it's worth.

14. If we decided to attempt driving the zebra mussels to local extinction, we would most likely use control programs that are density independent; the goal would be to make the greatest impact possible. (Density-independent control means that mortality is a set percentage of the population, not dependent on the population's density. For example, raising salinity in the lake could perhaps kill 90% of the zebra mussel population, whether it is a population of one billion or one trillion.) Using the data that you analyzed outside of class, identify which site and which time would be best suited to using a control agent that is density independent. Justify your choice.

15. Given all the evidence and natural history on the zebra mussel, put together an argument supporting the allocation of large funds for the control of zebra mussels. Use your answers to the above questions to aid your discussion. You should address the economic, as well as the ecological, impacts of the mussel. For economic impacts, think about industry, recreation, and tourism. For ecology, consider both the natural history background and the data provided to support your answer. Is the zebra mussel actually a pest? Are natural enemies present? If a control agent is to be used, would the environmental benefit be greater than the cost? Is the control practical?

References

Finkel, Elizabeth. 1999. Australian biocontrol beats rabbits but not rules. *Science* 285(5435): 1842.

Hager, H A., and K. D. McCoy. 1998. The implications of accepting untested hypotheses: a review of the effects of purple loosestrife (*Lythrum salicaria*) in North America. *Biodiversity and Conservation* 7: 1069–1079.

Lori, E., and S. Cianfanelli, 2006. New records of *Dreissena polymorpha* (Pallas, 1771) (Mollusca: Bivalva: Dreissenidae) from central Italy. *Aquatic Invasions* 1(4): 281–283.

Ludyanskiy, M. L., D. McDonald, and D. MacNeill, 1993. Impact of the zebra mussel, a bivalve invader. *Bioscience* 43(8): 533–544.

McKnight, Bill N. 1993. *Biological Pollution: The Control and Impact of Invasive Exotic Species.* Indianapolis, In.: Indiana Academy of Science.

Nature Conservancy. America's Least Wanted: Alien Species Invasions of U.S. Ecosystems. http://consci.tnc.org/library/pubs/dd/toc.html.

Ricciardi, A., and H. J. MacIsaac. 2000. Recent mass invasion of the North American Great Lakes by Ponto-Caspian species. *Trends in Evolutionary Ecology.* 15(2) 62–65.

Ross, J., and M. F. Sanders. 1984. The development of genetic resistance to myxomatosis in wild rabbits in Britain. *The Journal of Hygiene* 92: 255–261.

Schloesser, D., T. Nalepa, and G. Mackie. 1996. Zebra mussel infestation of unionid bivalves (Unionidae) in North America. *American Zoologist* 36: 300 –310.

Stone, Richard. 1999. Keeping paradise safe for the natives. *Science* 285(5435): 1837.

9 Island Biogeography: Evaluating Correlational Data and Testing Alternative Hypotheses

James Robertson

Introduction and Background

The earth's surface is spatially complex—even the most casual observer is readily aware of this. From the lush tropical zones of the equator to the desolate arctic zones of the poles, habitat diversity exists on both a global and local scale. Organisms live in almost every imaginable habitat on earth. However, no one species is able to persist in all of earth's habitats (aside from humans, perhaps). As you will remember from earlier chapters, the constraints of any environment mean that no single strategy is likely to be optimal in all environments. As natural selection acts and evolution occurs, diversity emerges. The interdisciplinary field of biogeography seeks to understand what influences the distribution of species. Biogeographers document the geographic distributions of species (both present and past), and try to understand the underlying processes that have led to these spatial patterns (Brown and Lomolino, 1998). As early as the eighteenth century, naturalists, precursors to today's biogeographers, attempted to record the geographic distributions of species, particularly species with discontinuous distributions. In some cases, discontinuous distributions can be explained by a dispersal event across an inhospitable habitat. However, when a number of species have the same discontinuous distribution, the most parsimonious explanation is the formation of a natural barrier separating geographic regions. Vicariance biogeography is the study of closely related taxa that are separated geographically as a result of the formation of natural barriers to dispersal. Many classic examples of vicariance biogeography make sense in light of plate tectonics. The earth's outermost layer is broken into large plates that float and shift on a molten underlying layer. In the late 1960s this theory (originally proposed by Antonio Snider-Pelligrini in 1858) became widely accepted and was used to explain global distributions of many plant and animal taxa. Before plate tectonics were understood, widely distributed species were assumed to have dispersed across vast distances to colonize hospitable regions.

Consider the distribution of ratites, large flightless birds (including emus and ostriches). They are found throughout the southern hemisphere. Their distribution was assumed to have been the result of multiple oceanic dispersal events. The theory of plate tectonics, however, provided a more parsimonious explanation: ratites occupied the Southern Hemisphere when South America, Africa, and Australia were still a single landmass. As plates shifted, these three landmasses drifted, and ratites became isolated on each continent. Eventually, as a result of isolation, possible founder effects, and local selective pressures, the ratites became today's African ostriches, South American rheas, and Australian emus

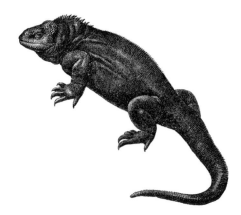

FIGURE 9.1. Land iguana, one of the Caribbean lizards.

and cassowaries. The theory of plate tectonics helps explain many such global species distributions (Brown and Lomolino, 1998).

Biogeographers are not only interested in understanding the origin and spread of species; they are also interested in understanding the factors responsible for geographic patterns in both the numbers and the types of species. For example, naturalists in the nineteenth century noted a general trend in the diversity of island species: as island size increases, the number of species present typically increases. This relationship can be expressed graphically with the species-area curve (Brown and Lomolino, 1998). Scholars originally focused on understanding the biogeography of (oceanic) islands, but today, species-area curves have been extended to include *functional* "islands": e.g., a series of ponds separated by land, or a series of mountain peaks separated by low-lying deserts. These are functional islands because organisms living there may be unable to move among isolated habitable areas across intervening inhospitable areas.

MacArthur and Wilson (1969) explained the species-area correlation as a balance between extinction and immigration. They suggested that larger islands will have greater habitat diversity and thus allow more species to persist (i.e., avoid extinction). The island's distance from the mainland also influences the immigration rate; more distant islands are more rarely colonized (figure 9.1 and table 9.1). The extension of the species-area curve to include functional islands is particularly interesting in today's world of rapid human population growth and urban sprawl. As human populations increase globally, natural habitats have become increasingly fragmented. These fragmented patches, which in most cases continue to decrease in size, serve as island habitats for many species of plants and animals. Conservation biologists have used species-area curves to predict extinction rates due to habitat loss.

The allometric equation was invented to understand how size (or biomass, or some other reflection of size) affects processes, from metabolic rates to species diversity. In the 1920s a Swedish scientist, Arrhenius, adapted the allometric equation to describe the species-area curve: $S = cA^z$. In this equation, S is the number of species present and A is the area of the island. Both c and z are fitted constants, with z setting the shape and c the placement of the curve. This relationship can be expressed linearly by taking the logarithm of both sides of the equation. This is a linear conversion of the relationship of island size and species number: z is the slope of that line, and the constant, $\log(c)$, is the y-intercept. To evaluate this model, we need data on both the area and the number of species on islands of interest (see table 9.1).

TABLE 9.1.
Herpetological biodiversity in the West Indies.

Island name (West Indies)	Area (km²)	Species no.	log (area (km²))	log (species no.)
Redona	2.6	3	0.4138	0.4771
Saba	12.7	5	1.104	0.699
Monteserrat	104	9	2.016	0.9542
Puerto Rico	8816	40	3.945	1.602
Jamaica	11669	39	4.067	1.591
Hispaniola	77793	84	4.891	1.924
Cuba	103723	76	5.016	1.881

Source: Darlington (1957).

$$S = cA^z,$$
$$\log(S) = \log(cA^z),$$
$$\log(S) = \log(c) + \log(A^z), \quad (9.1)$$
$$\log(S) = \log(c) + z\log(A).$$

The model requires that the c and z constants do not vary within a pooled data set. If they do, then the equation will be a poor predictor of the relationship between island size and species number. The question, then, is how to avoid pooling data with different c and z constants. We must be able to predict when the c and z constants will differ; to do this we must have a *biological* understanding of these constants. There has been a great deal of controversy concerning the biological implications of these two constants. The general view is that the value of c (which affects the placement) depends on the taxon and the biological region of the globe, and that z (which affects the shape of the curve) depends on the degree of island isolation.

This approach is valid only when you are comparing within a group with similar mobility and life history (usually taxonomically close groups), on environmentally similar islands, with similar degrees of isolation. For example, you do not want to pool species of ants and species of birds on the same graph, or to compare islands in the arctic tundra of Alaska with islands in the Caribbean, or to compare islands just off the Florida coast with the relatively isolated islands of Hawaii. Unfortunately, there is no standard method for determining what data are appropriate to pool—and as a result, pooling is often the result of an investigator's intuition. Problems of inappropriate data are common to many types of analysis, not just species-area curves. Always consider this potential problem when evaluating reported results (Brown and Lomolino, 1998).

Homework for this exercise takes approximately 45 minutes.

Objectives of This Exercise

In this exercise you will use the basic principles of island biogeography to examine data critically. You will calculate and compare coefficients of determination to decide how well the data fit island biogeography models.

Case Studies and Data

The relationship between island size and species number is very appealing to biogeographers. Fortuitously, when we began to think about island biogeography, extensive data already existed and could be used to test our hypotheses. As long ago as 1957, Philip J. Darlington, Jr. compiled data on the number of reptilian and amphibian species on islands in the West Indies in the Caribbean. What pattern do you expect when you graph the "raw" data? What if you plot the log of the area against the log of the species number, as in figure 9.2?

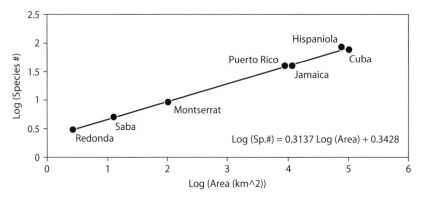

FIGURE 9.2. Species-area curve: Amphibians and reptiles of the West Indies.

If the allometric equation $S = cA^z$ correctly predicts the relationship between island size and species number, then we expect the log-log graph to be a neat straight line. We can see that the values do, in fact, lie along a straight line. The line on the graph is the least squares regression "line of best fit" (see chapter 14). In this case the data fit almost perfectly—for these data, the equation $S = cA^z$ closely predicts the relationship between island size and species number. Other data sets, however, may not fit so well, in large part depending on how effectively isolated the islands are, or how well island size predicts available ecological diversity.

We need a way to quantify how well the data fit a linear relationship. One standard method used to quantify how well the line describes the relationship is to calculate the proportion of the variation in the Y variable (here, the log of the number of species) that is explained by the X variable (here the log of island area). We call this R^2, the coefficient of determination.

$$R^2 = \frac{[n(\Sigma\ x_iy_i) - (\Sigma\ x_i)(\Sigma y_i)]^2}{[n(\Sigma\ x_i^2\ ) - (\Sigma\ x_i)^2]\ [n(\Sigma\ y_i^2) - (\Sigma\ y_i)^2]}.$$ (9.2)

See chapters 14 and 15 for more examples, details, and explanation of R^2.

Questions to Work on Individually Outside of Class

1. Examine the species area curve for West Indies reptiles and amphibians (figure 9.2); you can readily see that the line has a positive slope and that the R^2 will be close to 1. We still, however, need to calculate the actual value. Fill out table 9.2 so that you will be able to

TABLE 9.2.
Worksheet for calculating coefficient of determination for West Indies amphibian and reptile species-area curve.

Island name (West Indies)	log (area (km²)) (x)	[log (area (km²))]² (x)²	log (species no.) (y)	[log (species no.)]² (y)²	log (A) log (species no.) (x)*x(y)
Redona	0.4138	.1712	0.4771	.2276	.1974
Saba	1.1040		0.6990		
Montserrat	2.0160		0.9542		
Puerto Rico	3.9450		1.602		
Jamaica	4.0670		1.591		
Hispaniola	4.8910		1.924		
Cuba	5.0160		1.881		
Sum	21.4500				

calculate R^2 for the next question. Several cells have been calculated for you; ignore them initially and use them to check your calculations.

2. Calculate the coefficient of determination (R^2). SHOW YOUR WORK!!!

3. Would you have predicted this result simply by looking at the graph? If points were added to the graph that did not fall on or near the line what would happen to R^2?

Now, we will add two more data sets, and make specific comparisons across them.

FIGURE 9.3. Long-horned beetle.

Log (Sp.#) = 0.4496 Log (Area) + 0.803
R^2 = 0.5409

FIGURE 9.4. Species-area curve : Cerambycid beetle species of the Florida Keys (from Brown and Peck, 1996).

TABLE 9.3.
Long-horned beetle species numbers in the Florida Keys.

Island name (Florida keys)	Area (km^2)	Distance (km)	Species no.	log (distance)	log (species no.)
Key Largo	55.1	13	44	1.114	1.643
Matecumbe Key	4.3	32	16	1.505	1.204
Fat Deer Key	3.7	66	12	1.82	1.079
Key Vaca	2.9	72	15	1.857	1.176
No Name Key	3.1	76	16	1.881	1.204
Nig Pine Key	17.1	79	24	1.898	1.38
Big Torch Key	2.3	88	16	1.944	1.204
Cudjoe Key	9.2	95	8	1.978	0.903
Sugarloaf Key	10.2	100	10	2	1
Key West	11.9	121	24	2.083	1.38
Dry Tortugas	0.9	131	3	2.117	0.477

Source: Brown and Peck, 1996.

Cerambycid Beetles of the Florida Keys Browne and Peck (1996) sampled long-horned beetle species (Family: Cerambycidae; figure 9.3) in the Florida Keys to examine the relationship between island area and species number. They found 53 different species across 11 different islands, which varied both in size (from .9 to 55.1 km^2) and in distance (from 13 to 131 km) from the Florida coast. Table 9.3 includes some of their data. Figure 9.4 is a log-log plot of island area and species number. Included in the graph is the equation for the best-fit line, along with R^2.

Land Birds of the West Indies Terborgh's (1973) data on the relationship between island area and species number for bird species in the West Indies are found in figure 9.5.

Across 19 islands (varying in size from 83 to 114,521 km^2) there are finches, warblers, thrushes, mimic thrushes, hummingbirds, pigeons, flycatchers, woodpeckers, owls, hawks, and falcons. Examine the log-log plot of island area and species number. Included in the graph is the equation for the best-fit line, along with R^2.

Examine the map of the Caribbean Sea including islands of the West Indies and the Florida Keys (figure 9.6).

4. Complete table 9.4 so that you can compare the data on the different taxa.

TABLE 9.4.
Comparison of MacArthur and Wilson's model across taxa.

Taxon group	Location	log (c)	c	z	R^2
Amphibians/reptiles	West Indies				
Beetles	Florida Keys				
Birds	West Indies				

FIGURE 9.5. Species-Area Curve: Land Birds of the West Indies (from Terborgh, 1973).

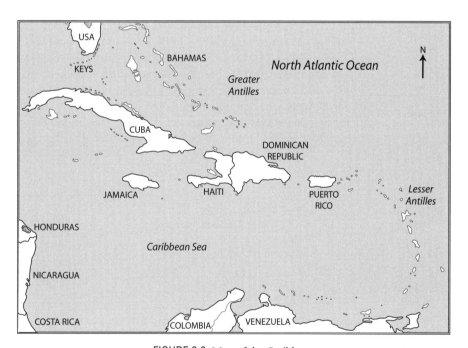

FIGURE 9.6. Map of the Caribbean.

Would pooling the three data sets (reptiles and amphibians, beetles, and birds) be appropriate? Why or why not? You should consider all aspects of the biological interpretation of these constants in your answer. You may also want to consult the Caribbean map.

5. How well does each of the three data sets support the relationship predicted by the equation $S = cA^z$? Assume that R^2 above 0.85 is excellent, between 0.70 and 0.85 is good, and below 0.70 is poor. Provide both a biological and a statistical explanation for these results.

Small-Group/In Class Exercise

Understanding Biogeography of Channel Island Birds

The allometric equation $S = cA^Z$, introduced by Arrhenius in the 1920s, predicted a specific and testable relationship between island size and species number. Biogeographers have examined this relationship, expressed graphically as the species-area curve, in a variety of taxonomic groups and in extremely diverse island habitats. Some of these data sets, such as Darlington's (1957) data on reptiles and amphibians of the West Indies, strongly support the predicted relationship between island area and species number. Many other data sets, however, have low coefficients of determination (R^2), suggesting that perhaps a more complex relationship between species number and island area exists (Brown and Lomolino, 1998; Spellerberg and Sawyer, 1999). Motivated in part by the inconsistencies in the species-area curve across examples, Robert H. MacArthur and Edward O. Wilson (1967) examined the relationships among island size, degree of isolation, and species number. Both men were very familiar with island systems. MacArthur had studied bird species on montane "islands" of the southwestern United States, islands of the West Indies, and islands off the coast of Maine and Panama. Wilson had studied ant species on islands of the East Indies, Polynesia and the Florida Keys. In their seminal work, *The Theory of Island Biography* (1967), MacArthur and Wilson incorporated many preexisting concepts in ecology, population biology, and biogeography into a single island biogeographic model (Brown and Lomolino, 1998; Spellerberg and Sawyer, 1999). Figure 9.7 depicts the relationship between species number and both the rate of colonization and the rate of extinction for a particular island. The intersection of the colonization rate and the extinction rate represents the species number at equilibrium as indicated on the *x*-axis. The point *P* located

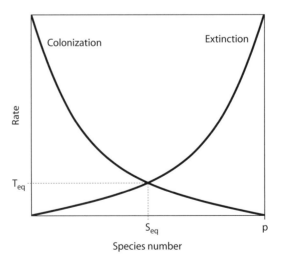

FIGURE 9.7. The relationship between species number and both the rate of colonization and the rate of extinction for a particular island. The intersection of the colonization rate and the extinction rate represents the species number at equilibrium as indicated on the *x*-axis. The point *P* located on the *x*-axis indicates the number of species in the source pool or on the mainland. The species turnover rate occurs at the *y*-intercept of the intersection point (adapted from MacArthur and Wilson, 1967).

on the *x*-axis indicates the number of species in the source pool (e.g., on the mainland). The species turnover rate occurs at the *y*-intercept of the intersection point (adapted from MacArthur and Wilson, 1967).

In MacArthur and Wilson's model, the equilibrium number of species on an island depends on the balance between rate of colonization and the rate of extinction. MacArthur and Wilson argued that the colonization rate should decrease as species number increases. Initially, any species reaching the island would be new to the island and therefore represent a colonization event. Eventually, the colonization rate should decrease until all the species found on the mainland are also found on the island. If a species goes locally extinct on the island, then that species may again eventually recolonize the island, contributing again to the colonization rate. The extinction rate, however, should increase as species number increases. Initially, extinction rates are zero because there are no species to go extinct. However, as the number of species on the island increases, the number of species that can potentially go extinct will also increase, thereby increasing the possible extinction rate. In theory the balance between these two rates produces a stable equilibrium species number, (the *x*-coordinate of the intersection point). Interestingly, although the number of species on the island should remain the same over time, the actual species represented on the island should change continuously. This is a consequence of the ongoing processes of extinction and recolonization that occur even at the equilibrium point. The rate at which species are replaced on the island when the number of species is at the equilibrium species number is the species turnover rate (the *y*-coordinate of the intersection point) (figure 9.8).

The concept of a species turnover rate was critical to the MacArthur and Wilson model. Biogeographers had previously assumed that so long as the island environment remained relatively constant, the particular species on the island would not change. MacArthur and Wilson, however, suggested that even if the island environment remained constant the particular species present on the island would continue to change (Brown and Lomolino, 1998; MacArthur and Wilson, 1967; Spellerberg and Sawyer, 1999). MacArthur and Wilson also considered how island area and degree of isolation would influence the rate of colonization and the rate of extinction. They suggested that colonization rates depended largely on the degree of island isolation from a mainland source, decreasing with isolation. They noted that increases in island size might increase colonization rates because of the increased likelihood of immigrants finding the larger island (referred to as the target effect). However, they argued that the influence of island size on colonization rates would be insignificant when considered along with the influence of isolation (MacArthur and Wilson, 1967).

Extinction rates, MacArthur and Wilson suggested, depend largely upon population size. As the population size decreases, we see increasing influence of stochastic processes that can lead to local extinction. For example, a few random deaths in a very small population will have a greater impact then the same number of deaths in a large population. Populations of small size are also more likely to be affected by inbreeding depression (an increased rate of homozygosity of deleterious recessive alleles due to mating between closely related individuals; see chapter 12 for detailed examples). In some cases this can also lead to local extinction (Lincoln et al., 1998). Population size should increase as island area increases, as a result of an increase in available resources. Therefore, extinction rates should decrease with increasing island size. The balance between colonization rate (dependent on the degree of island isolation) and extinction rate (dependent on island size) determines the number of species at equilibrium and the species turnover

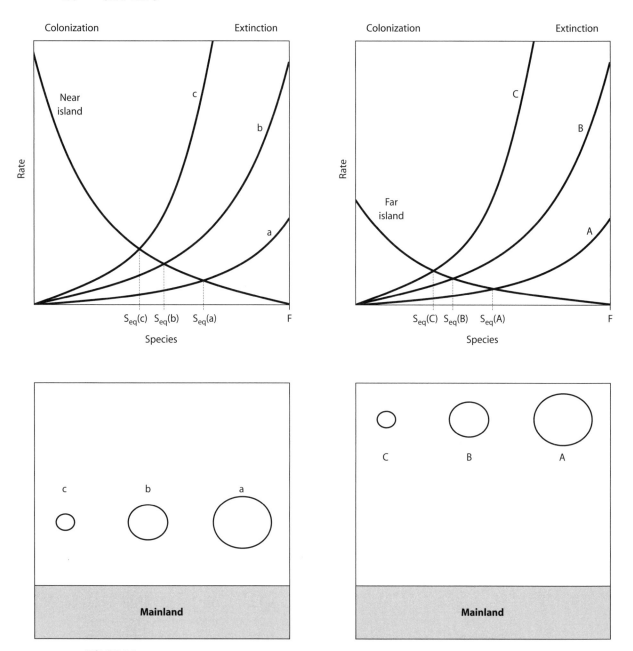

FIGURE 9.8. Both the rate of colonization and the rate of extinction vary for islands of different size and distance from the mainland or source population. The predicted equilibrium species number is indicated on the x-axis (adapted from MacArthur and Wilson, 1967).

rate (Brown and Lomolino, 1998; MacArthur and Wilson, 1967; Spellerberg and Sawyer, 1999).

MacArthur and Wilson did not consider the influence of island isolation on extinction rates, but extinction rates may be influenced by the degree of island isolation. For example, if an island is relatively close to a mainland source, or to other islands, then a constant flow of new immigrants can maintain even small populations and thereby reduce the frequency of local extinction (referred to as the rescue effect).

Case Study

Jared Diamond (1969) examined the basic principles of the island biogeography model in bird species (both land and freshwater) breeding on nine of the Channel Islands off southern California. Diamond chose this study system because bird species had been surveyed on these same islands in 1917, 51 years prior, providing a comparison for examining one of the fundamental concepts of MacArthur and Wilson's model: the species turnover rate (figure 9.9)

Part A: Written Study Proposal

You will now have a chance to apply what you have learned about species-area curves and various aspects of the MacArthur and Wilson island biogeography model. As a group, you will develop a written proposal to test the various predictions made in the island biogeography model with data on Channel Island bird species. Some of your analyses may simply involve comparisons between columns; however, other analyses may require log transformation and evaluation of correlation coefficients. You may find it necessary to test ideas not directly addressed in the models; however, if you do so, you must justify

FIGURE 9.9. Channel Islands.

TABLE 9.5.
Channel Island birds species tallies.

Island name	Area (km^2)	Distance (km)	Species no. (1968)	No. extinct	Human introduced	Colonization	Turnover (%)
Los Coronados	2.6	13	11	4	0	4	36
San Nicholas	57	98	11	6	2	4	50
San Clemente	145	79	28	9	1	4	25
Santa Catalina	194	32	30	6	1	9	24
Santa Barbara	2.6	61	10	7	0	3	62
San Miguel	36	42	11	4	0	8	46
Santa Rosa	218	44	14	1	1	11	32
Santa Cruz	249	32	36	6	1	6	17
Anacapa	2.9	21	15	5	0	4	31

your reasons for including these analyses. If the model predicts that two variables are not correlated, you may wish to test the relationship between those variables. You may test only data included in table 9.5. For example, you have no data on environmental variation between islands, and therefore you cannot test how this may influence species number.

Write a brief proposal to test predictions of the models with data from the Channel Islands. Your proposal should address each of the following:

- What aspect of MacArthur and Wilson's model are you testing?
- How would you analyze the data?
- What do you expect to see if the data support the model, and what do you expect to see if the data do not support the model?
- What assumptions must you make when analyzing the data?
- What other information do you wish we had collected on these islands?

After you have completed your written proposal, your instructor will give you, when possible, the graphical and statistical analyses that you requested. You will then be able to evaluate the data and present your conclusion to the class.

Part B: Oral Presentation of Analytical Results

Carefully examine the analyses provided by your instructor. As a group, you can now prepare an oral presentation discussing your prediction, the analysis, and your conclusions. You must specifically address whether the data support your analysis of the particular aspects of the model you chose to examine. You should clearly outline all of your assumptions and any limitations of the analysis you are presenting.

Finally, consider the following criticism of Diamond's work and outline further work that could clarify our understanding of Channel Island bird biogeography.

In his original paper Diamond (1968) pointed out that the extinction rates, as well as the turnover percentage, have probably been grossly underestimated. This is because multiple species may have colonized the islands and then gone extinct between the two survey times; they would not have been included in the data. Species present during either or both of the surveys may also have gone extinct and recolonized the islands any number of times between the two surveys. Based on this, Diamond concluded that the actual extinction rate and subsequent turnover percentage would most likely be higher than the values estimated in table 9.5.

Lynch and Johnson (1974) pointed out that many of the island extinctions could be attributed to human influences. For example, the disappearance of birds of prey like the Osprey, Bald Eagle, and Peregrine Falcon, which accounted for a large portion of the extinctions, was most likely due to pesticide poisoning and habitat modification. They also argued that much of the colonization was the result of the immigration of house sparrows and European starlings. Both of these species were introduced from Europe and are considered invasive (introduced species capable of significantly increasing their distribution from the point of introduction, and in the process out-competing native, endemic species with similar ecological requirements). In such cases the endemics may be driven to extinction (Brown and Lomolino, 1998, Lynch and Johnson, 1974).

Diamond continued to monitor bird species on Santa Catalina for several years after the study. He found year-to-year variation in the particular species breeding on the island. However, in most cases, only highly migratory birds varied in their presence or absence on the island. Even a distance of several kilometers would most likely have little meaning to these birds (Brown and Lomolino, 1998; Jones and Diamond, 1976).

Your group will have five minutes to explain your prediction, analysis, and interpretation of the results provided by the instructor. In addition, be sure to take a minute to explain further work you might want to perform to further test the equilibrium model and species area predictions for the Channel Islands.

References

Brown, J. H., and M. V. Lomolino. 1998. *Biogeography*. Sunderland, Mass.: Sinauer Associates.

Browne, J., and S. B. Peck. 1996. The long-horned beetles of south Florida (Cerambycidae: Coleoptera): Biogeography and relationships with the Bahama Islands and Cuba. *Journal of Zoology* 74: 2154–2169.

Darlington, P. J., Jr., 1957. *Zoogeography: The Geographic Distribution of Animals*. New York: John Wiley & Sons.

Diamond, J. M. 1968. Avifaunal equilibrium and species turnover rates on the Channel Islands of California. *Proceedings of the National Academy of Science* 64: 57–63.

Jones, H. L., and J. M. Diamond. 1976. Short-time-base studies of turnover in breeding bird populations on the California Channel Islands. *Condor* 78: 536–549.

Khazanie, R. 1990. *Elementary Statistics in a World of Applications*. New York: Harper Collins.

Lincoln, R., G. Boxshall, and P. Clark. 1998. *A Dictionary of Ecology, Evolution and Systematics*. Cambridge: Cambridge University Press.

Lynch, J. D., and N. V. Johnson. 1974. Turnover and equilibria in insular avifaunas, with special reference to the California Channel Islands. *Condor* 76: 370–384.

MacArthur, R. H., and E. O. Wilson. 1967. *The Theory of Island Biogeography.* Princeton, N.J.: Princeton University Press.

Spellerberg, I. F., and J.W.D. Sawyer. 1999. *An Introduction to Applied Biogeography.* Cambridge: Cambridge University Press.

Terborgh, J. 1973. Chance, habitat and dispersal in the distribution of birds in the West Indies. *Evolution* 27: 338–349.

Section III

Population Genetics

10 Hardy-Weinberg: Evaluating Disequilibrium Forces

Jason J. Kolbe

Introduction and Background

Human modification of the environment fragments habitats around the world. Just as oceanic islands are isolated by water, terrestrial habitat islands become isolated by unsuitable or inhospitable habitats (see chapter 9 for more). Habitats can become fragmented as roads and buildings are built, forests are clear-cut, or wetlands are drained. Habitat fragmentations and the consequential isolation of small populations are among the most critical contemporary conservation issues. Evolutionary processes at the genetic level (e.g., inbreeding, genetic drift, and gene flow) can lead to reduced genetic variation, which can further disrupt the potential for adaptive evolutionary change in small isolated populations. The Hardy-Weinberg equilibrium theory (H-W) is a fundamental concept in population genetics that provides the foundation for much of conservation genetics. Hardy-Weinberg establishes the null hypothesis for studies of allele frequency change, or microevolution, in populations. According to H-W, allele and genotype frequencies should remain constant over successive generations when the following circumstances are met: (1) large population size, (2) no migration across populations, (3) no net accumulation of mutations, (4) random mating, and (5) no natural selection. As you might guess, these are stringent conditions, rarely met.

When we find departures from expected allele and genotype frequencies, we must reject the null hypothesis (see also chapter 15). If the null hypothesis is rejected, the population is in disequilibrium; we can look for the sources of change in allele or genotype frequencies. Five potential disequilibrium forces include: genetic drift, gene flow, mutation, nonrandom mating, and natural selection (table 10.1). Compare these processes to the conditions required for H-W to predict equilibrium.

Of the (often anthropogenic) potential causes of microevolutionary change, the most important for conservation are genetic drift, gene flow across small populations, and nonrandom mating in small isolated populations. How many migrants do populations lose, receive, or exchange per generation? Has a past population bottleneck resulted in loss of heterozygosity from genetic drift, or inbreeding? These (and other important conservation questions) can be addressed using population genetic techniques based on the H-W theory. A working knowledge of this basic population genetic principle is valuable for conservation biologists.

Two fundamental equations are basic to H-W theory. First, for one gene locus with two alleles (i.e., alternative forms of the same gene on different chromosomes), the sum

TABLE 10.1.
Assumptions, processes, and conservation issues relevant to H-W theory.

Violations of H-W equilibrium	Processes in nature	Relevant conservation issues
Genetic drift	Stochastic events Bottlenecks Founder events	Habitat fragmentation reducing population size
Gene flow	Migration	Habitat fragmentation increasing distance among subpopulations
Nonrandom mating	Inbreeding Assortive mating	Reduced population size and mating preferences
Natural selection	Differential survival	Introduced predators and pesticides; adaptation and reproduction
Mutation	DNA sequence changes	Source of raw genetic material at a rate of 1 mutation per 10^5 or 10^6 years

of the two allele frequencies must equal 1: $p + q = 1$, where the frequency of one allele is p and the other is q. When one allele is known, the other can always be calculated: $p = 1 - q$ or $q = 1 - p$. For example, imagine that the tail fur of a mouse is controlled by a single gene with two alleles H_1 = hairy and H_2 = hairless. If we know that 40% of the alleles for tail fur in a population are H_1, then we know that 60% of the alleles in that population are H_2.

Second, the genotype frequencies for homozygous ($H_1H_1 = p^2$ and $H_2H_2 = q^2$) and heterozygous ($H_1H_2 + H_2H_1 = 2pq$) genotypes must equal 1 as well: $p^2 + 2pq + q^2 = 1$. If the population of mice discussed above are in Hardy-Weinberg equilibrium, the homozygous H_1H_1 frequency is expected to be 16% ($p^2 = [0.40]^2 = 0.16$); the heterozygous H_1H_2 frequency is expected to be 48% ($2pq = 2[0.40][0.60] = 0.48$); and the homozygous H_2H_2 frequency is expected to be 36% ($q^2 = [0.60]^2 = 0.36$). As required, the sum of the genotype frequencies is 1 ($0.16 + 0.48 + 0.36 = 1$). If genotype frequencies are known, allele frequencies can be calculated, as follows: p = frequency of H_1H_1 + 1/2 frequency of H_1H_2 and q = frequency of H_2H_2 + 1/2 frequency of H_1H_2.

These fundamental equations provide the basic tools for assessing whether a population's allele or genotype frequencies are in Hardy-Weinberg equilibrium, whether one (or more) of the assumptions have been violated, and whether microevolutionary change is occurring.

Objectives of This Exercise

In this exercise you will apply Hardy-Weinberg equilibrium theory to problems in conservation genetics. You will analyze population genetic data and write a report discussing the results of the project focusing on conservation recommendations derived from the population genetic analysis. Although collared lizards are real, the data you will use in this exercise were generated to illustrate the model.

Case Study and Data

You and your colleagues have been awarded a grant from the Missouri Department of Conservation and The Nature Conservancy to determine the genetic consequences of habitat fragmentation in the Missouri Ozarks. The Ozarks of southern Missouri and northern Arkansas are a region of great geological and topographical complexity, with a rich history of floral and faunal change associated with climatic fluctuations since the Pleistocene era (~12,000 years ago). Many unique plants and animals live in remnant habitats, called glades, which are more characteristic of prairies and deserts than of the surrounding oak-hickory forests. Glade populations are of conservation concern because the encroaching forests both reduce the size of individual glades and increase the distance between glades. Fire suppression has facilitated this habitat fragmentation in the Ozarks, by allowing forest species to become established in areas where natural fires previously excluded them. The demographic and genetic viability of many plant and animal populations restricted to these glade habitats is uncertain.

You have been assigned to work on the charismatic collared lizard, *Crotaphytus collaris* (figure 10.1) a species with a range from the desert southwest, through the Great Plains, to the Missouri Ozarks (figure 10.2). Collared lizards are habitat-specific: they require open

FIGURE 10.1. Collared lizards *Crotaphytus collaris.*

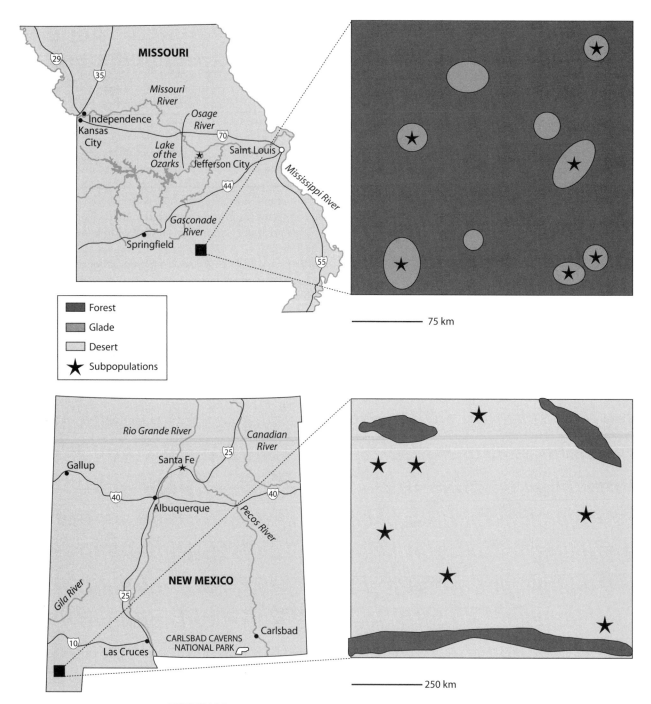

FIGURE 10.2. Maps of the Missouri Ozark and New Mexico populations of collared lizards *Crotaphytus collaris*.

deserts, prairies, or glades, and are not found in forests, woodlands, or disturbed areas. The collared lizard's range is continuous in the desert southwest and Great Plains, but highly fragmented in the Ozarks because of the lizard's restriction to glades.

An explicit requirement of the grant is that you compare the genetic structure of fragmented populations in the Ozarks to unfragmented populations in other parts of the species' range. This approach will allow you to determine the consequences of habitat fragmentation on the collared lizard's population genetic structure.

Questions

In this exercise, you will track the allelic and genotype frequencies across three generations for two collared lizard populations: one from New Mexico, the other from the Missouri Ozarks. The New Mexico population is located in Chihuahuan Desert National Park (CDNP), a 1,000-hectare park; it is part of a diverse reptile community. The Ozark population is located on private and state lands ranging in size from 0.5 to 15 hectares; this population is one of only a few lizard species found in glades. The desert habitat in CDNP is continuous, but the glade habitat in the Ozarks is highly fragmented by intervening forest (figure 10.2). Ozark subpopulations are separated by 5 to 290 km. Male collared lizards have three distinct ventral coloration phenotypes: orange, yellow, and blue. This coloration is important in their mating behavior, and follows a simple pattern of inheritance. If a male is homozygous for the A_1 allele ($A_1 A_1$), its belly is orange; if it is heterozygous ($A_1 A_2$), the belly is yellow; and if it is homozygous for the A_2 allele ($A_2 A_2$), its belly is blue. Data on the number of individuals in each phenotype for both the New Mexico and Ozark populations are given in tables 10.2 and 10.3).

Calculate the observed genotype frequencies, allele frequencies, and expected genotype frequencies for each of the three generations; determine whether the population is in Hardy-Weinberg equilibrium in each generation, by comparing the observed and expected genotype frequencies.

Questions to Work on Individually Outside of Class

1. Complete the allele and genotype frequency calculations and fill in the blanks in tables 10.2 and 10.3 for the New Mexico and Ozark populations of collared lizards, for all three generations.
2. Then,

(a) Are the New Mexico and Ozark populations of collared lizards in H-W equilibrium? If observed and expected genotype frequencies are within 0.05 (this is not a p-value, but a proportion), then consider the populations to be in H-W equilibrium.

(b) If the populations are not in H-W equilibrium, which single disequilibrium force is the most likely explanation for the rejection of the H-W equilibrium in these populations?

(c) We often do not have the luxury of 3 generations of data on allele frequencies. Could you determine whether these populations are in H-W equilibrium with only one generation of data?

3. Discuss each of the five mechanisms listed in table 10.1 in light of their potential effects on both the New Mexico and Ozark populations of collared lizards. Use all the information in the tables, including the trends observed across three generations. Are some

TABLE 10.2.
Collared lizards in the New Mexico region.

Generation 1 Phenotype	Orange	Yellow	Blue	Total
Genotype	A_1A_1	A_1A_2	A_2A_2	
Number of individuals	551	988	361	= 1900
Observed genotype frequencies	____	____	____	= 1.0
Allele frequencies		$p =$ ____	$q =$ ____	= 1.0
Expected genotype frequencies	____	____	____	= 1.0
Are these lizards in Hardy-Weinberg equilibrium (yes/no)?				
Generation 2 Phenotype	Orange	Yellow	Blue	Total
Genotype	A_1A_1	A_1A_2	A_2A_2	
Number of individuals	510	901	289	= 1700
Observed genotype frequencies	____	____	____	= 1.0
Allele frequencies		$p =$ ____	$q =$ ____	= 1.0
Expected genotype frequencies	____	____	____	= 1.0
Are these lizards in Hardy-Weinberg equilibrium (yes/no)?				
Generation 3 Phenotype	Orange	Yellow	Blue	Total
Genotype	A_1A_1	A_1A_2	A_2A_2	
Number of individuals	546	1053	351	= 1950
Observed genotype frequencies	____	____	____	= 1.0
Allele frequencies		$p =$ ____	$q =$ ____	= 1.0
Expected genotype frequencies	____	____	____	= 1.0
Are these lizards in Hardy-Weinberg equilibrium (yes/no)?				

disequilibrium forces more likely in the Ozark population than the New Mexico population? Why? What other information about these populations of lizards would be helpful to identify potential disequilibrium forces?

Small Group/ In Class Exercise

Fragmentation of Collared Lizard Population

Now that you have assessed whether the New Mexico and Ozark populations of the collared lizard are in H-W equilibrium, the next step is to determine to what extent the subpopulations in each region are fragmented. Fragmentation can be investigated in

TABLE 10.3.
Collared lizards in the Ozark region.

Generation 1 Phenotype	Orange	Yellow	Blue	Total
Genotype	A_1A_1	A_1A_2	A_2A_2	
Number of individuals	34	29	27	= 90
Observed genotype frequencies	____	____	____	= 1.0
Allele frequencies		$p =$ ____	$q =$ ____	= 1.0
Expected genotype frequencies	____	____	____	= 1.0
Are these lizards in Hardy-Weinberg equilibrium (yes/no)?				

Generation 2 Phenotype	Orange	Yellow	Blue	Total
Genotype	A_1A_1	A_1A_2	A_2A_2	
Number of individuals	23	18	21	= 62
Observed genotype frequencies	____	____	____	= 1.0
Allele frequencies		$p =$ ____	$q =$ ____	= 1.0
Expected genotype frequencies	____	____	____	= 1.0
Are these lizards in Hardy-Weinberg equilibrium (yes/no)?				

Generation 3 Phenotype	Orange	Yellow	Blue	Total
Genotype	A_1A_1	A_1A_2	A_2A_2	
Number of individuals	24	16	34	= 74
Observed genotype frequencies	____	____	____	= 1.0
Allele frequencies		$p =$ ____	$q =$ ____	= 1.0
Expected genotype frequencies	____	____	____	= 1.0
Are these lizards in Hardy-Weinberg equilibrium (yes/no)?				

two ways. First, you could conduct a mark-recapture study to determine the extent of migration among the subpopulations and how far individuals migrate. This is a very labor-intensive process; it might take many years of data collection to determine accurate rates of migration among the subpopulations. Further, even if you can accurately measure dispersal among subpopulations, you cannot guarantee that these dispersal events result in reproduction and gene flow, which is ultimately the currency of a population geneticist. Alternatively, the effects of fragmentation can be addressed at the genetic level. Historical processes and past connectivity among subpopulations leave an imprint in the genetic structure of populations through processes like gene flow and genetic drift. You can perform a genetic study to determine the amount of subdivision among populations. F-statistics are used to make comparisons of the amount of variation at different levels of organization within a population. The levels of structure considered are the individual (I),

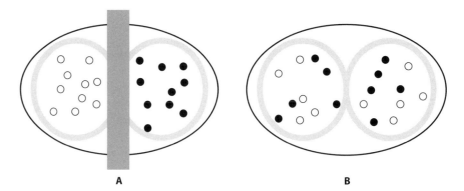

FIGURE 10.3. These two hypothetical populations, A and B, each have two subpopulations. Population A has a dispersal barrier that prevents migration between subpopulations. No dispersal barrier exists in population B and the subpopulations exchange migrants freely. Population A is completely subdivided ($F_{ST} = 1$), while population B is fully intermixed ($F_{ST} = 0$).

subpopulation (S), and total population (T). For example, a comparison of the amount of heterozygosity within an individual (I) relative to its subpopulation (S), is F_{IS}. This measure gives an estimate of the amount of inbreeding for that individual. F-statistics are also an effective way to calculate the amount of subdivision among populations. This is done by calculating F_{ST}, the ratio of the amount of heterozygosity within subpopulations (S) relative to all the populations combined (T). F_{ST} helps estimate the level of subdivision among populations. It ranges from 0 (no subdivision or a fully intermixing population) to 1 (complete subdivision or no migration) (figure 10.3). Thus, F_{ST} is directly related to the calculation of heterozygote frequencies that reflect whether a population is in H-W equilibrium (see table 10.1).

To assess the genetic structure in collared lizard populations from New Mexico and the Ozarks, you performed a microsatellite analysis, using the highly variable *wustl* locus, for seven subpopulations in New Mexico and six subpopulations in the Ozarks (figure 10.2). Based on heterozygote frequencies at this locus, you calculated F_{ST} values, We also know geographic distances for all possible pairs of subpopulations within the two regions (table 10.4). Looking at the relationship between F_{ST} and geographic distance will give you some insight into the historical connections among these subpopulations and the effect of their population sizes.

4. Plot F_{ST} versus geographic distance for all pairwise comparisons, in both the New Mexico and Ozark populations. Put geographic distance on the x-axis and F_{ST} on the y-axis. Draw a trend line and label the axes clearly. (Plot the New Mexico populations and the Ozark populations separately.)

5. Is there any relationship between F_{ST} and geographic distance? Is this relationship the same for the New Mexico and Ozark populations?

6. What do your plots suggest about the amount of subdivision, the amount of gene flow or migration, among the subpopulations for both the New Mexico and Ozark regions?

7. One way to estimate F_{ST} is from the relationship $F_{ST} = 1/(4N_{em} + 1)$. In this equation, the effective number of migrants is represented by N_{em}. Rearrange this equation and calculate the effective number of migrants per generation (N_{em}) for New Mexico subpopulation pairs 1-3, 1-7, and 4-7, and Ozark subpopulation pairs 2-3, 3-6, and 4-5. Describe the relationship between F_{ST} and N_em for these cases.

TABLE 10.4.

Pairwise comparisons of F_{ST} and geographic distances for seven subpopulations from New Mexico and six subpopulations from the Missouri Ozarks.

New Mexico			Ozarks		
Subpopulation comparison	F_{ST}	Geographic distance (km)	Subpopulation comparison	F_{ST}	Geographic distance (km)
1-2	0.14	280	1-2	0.36	90
1-3	0.03	20	1-3	0.70	5
1-4	0.09	70	1-4	0.26	50
1-5	0.11	750	1-5	0.57	180
1-6	0.09	540	1-6	0.35	220
1-7	0.18	560	2-3	0.43	150
2-3	0.05	390	2-4	0.80	240
2-4	0.03	220	2-5	0.50	45
2-5	0.07	260	2-6	0.25	290
2-6	0.15	760	3-4	0.89	160
2-7	0.09	310	3-5	0.62	200
3-4	0.08	80	3-6	0.72	270
3-5	0.04	130	4-5	0.58	10
3-6	0.16	520	4-6	0.83	110
3-7	0.14	570	5-6	0.62	80
4-5	0.01	10			
4-6	0.19	590			
4-7	0.33	910			
5-6	0.20	830			
5-7	0.18	460			
6-7	0.30	990			

8. Write a short report to the Missouri Department of Conservation and The Nature Conservancy outlining the general status of collared lizard populations in the Ozarks, compared to the status of New Mexico populations. Give your suggestions for managing the Ozark populations. Use all the information available to you, including the H-W data, the plots and their descriptions, and the background information on the species' biology and habitats. (One report per group.)

References

Campbell, N. A. 1996. *Biology*. 4th ed., San Francisco: Benjamin/Cummings.

Hartl, D. L. 1988. *A Primer of Population Genetics*, 2nd ed., Sunderland, Mass.: Sinauer Associates.

Hutchison, D. W., and A. R. Templeton. 1999. Correlation of pairwise genetic and geographic distance measures: inferring the relative influences of gene flow and drift on the distribution of genetic variability. *Evolution* 53: 1898–1914.

Meffe, G. K., and C. R. Carroll. 1997. *Principles of Conservation Biology*. Sunderland, Mass.: Sinauer Associates.

Templeton, A. R., K. Shaw, E. Routman, and S. K. Davis. 1990. The genetic consequences of habitat fragmentation. *Annals of the Missouri Botanical Gardens* 77: 13–27.

Templeton, A. R., R. J. Robertson, J. Brisson, and J. Strasburg. 2001. Disrupting evolutionary processes: the effect of habitat fragmentation on collared lizards in the Missouri Ozarks. *Proceedings of the National Academy of Sciences* 98: 5426–5432.

11 Drift, Demographic Stochasticity, and Extinction in Woggles

James Robertson, Anton Weisstein, and Stanton Braude

Introduction and Background

The forces that drive evolution have different impacts in large and small populations. In small populations, random events affecting survival and reproduction can have a great influence on overall frequencies of genes in a population and on the rate of extinction. This is far less true in large populations. Essentially this is a "sample effect" called genetic drift. In very small populations, drift can easily lead to the loss of alleles or even the extinction of a population. Variation in demographic parameters (e.g., sex ratio, age structure, age-specific fertility, and life span) may also have more profound effects on small populations than on larger ones. For example, in some species, when populations become extremely small the reproductive rate declines (this is one side of what is referred to as the Allee effect). However, any species is at greater risk of extinction when its population size is small.

When birth and death rates are equal, we expect a population to have a growth rate of zero and a stable population size (see chapter 6). However, *normal fluctuations in population size* can lead to extinction in small populations. For example, you would expect a population to increase if it has a birth rate of 0.6 births per female per year and a death rate of only 0.5 per female per year. However, every death is not immediately followed by a birth. In small populations, a run of deaths can quickly reduce the number of reproductive individuals and lead the population to extinction (Gotelli, 1998).

Random events affect small and large populations differently; we can see this by considering the chances of a tossed coin landing with heads or tails up. The probability of heads or tails on any given toss is 50% (0.5). The probabilities for landing heads up are constant, but if your sample of tosses is small (e.g., 10) you can sometimes get a high proportion of heads—even all heads. But in a large sample of tosses, the chance of huge or tiny proportions of heads is very small. Try this at home with coin tosses; of 20 ten-toss series, how many ten-toss games come up with 6, 7, or even 8 heads? If you repeated the game of ten-coin-tosses many times you could collect data on how often you get 7, 8, or 9 heads in ten tosses. You would be running a simulation that illustrates the variance around the predicted 5-head/5-tail average. Your simulation demonstrates that random fluctuations can potentially have a greater effect in small populations than in large populations.

Simulations can be valuable in predicting the fates of wild populations. Knowing the natural history and demography of a species helps us predict what is likely to happen to large populations. However, simulations allow us to investigate the effect of sampling and variance around average population parameters in small, versus large, populations. Of course we need to run any simulation many times to see the variation in possible outcomes.

In this exercise we will be using streamlined, simplified data on a fictional species, the woggle, to demonstrate how simulation models work, and to examine the effects of small population size on the probability of drift and extinction. Because life-history events rarely have probabilities that lend themselves to rolls of a die, we contrived a fictional rodent to illustrate how simple simulations work. *Woggles* are somewhat like kangaroo rats, but rather than run the risk that someone will imagine that our fictional data concern a real species, we have created an entirely fictional species. You will run an additional simulation with the woggles in chapter 13.

Homework for this exercise takes approximately one hour.

Objectives of This Exercise

This exercise shows (1) how population models and simulation data help us make predictions about the future survival of wild populations; and (2) how random events affect loss of genetic variation and extinction probability.

Case Study and Data

A recent survey of central California's desert region resulted in the discovery of a new rodent species. This particular rodent was placed in the polyphyletic family Cuteidae, and was given the Latin name *Treborus treborsonii*; its common name is "woggle" (see also figure 11.1 and chapter 13). Woggles have keen hearing; they are granivorous, bipedal, and saltatory (i.e., they are jumping grain-eaters). They have typical rodent countershaded coloration (dark on the dorsal side and white on the ventral side). Initial investigations showed that woggle populations, like lemmings, experience extreme fluctuations in population size over short periods of time. Over 30 populations have been sampled. Estimates of population size have varied greatly, from one population estimated to be about 300 individuals, to several populations possibly as small as 3 individuals The long-term viability of the very small populations is in doubt. Most of the populations experience extreme bottlenecks at some point, and many of the populations eventually go extinct. Recently, scholars have been concerned about the influence of stochastic processes on genetic variability and local extinction.

The maternally inherited mitochondrial (mt) genome has been sampled for over 500 individuals from 30 different populations; 25 unique mt genotypes are known. These various mt genotypes are named A, B, C, D, ... ,Y; each genotype is represented by a different letter. You will recall that the mt genome is maternally inherited, meaning that the mother passes her copy of the mtDNA to all her offspring, and the father does not contribute to the mtDNA. Both female and male offspring, however, inherit mtDNA.

FIGURE 11.1. The mythical woggle *Treborus treborsonii.*

At this moment, no one knows possible selective advantages of one genotype over another, so we will begin the exercise by assuming that all genotypes are selectively neutral.

Extremely harsh environmental conditions, including heavy predation by raptors and snakes, and extreme (and unpredictable—see chapter 18) fluctuations in weather conditions, have resulted in an average life span for woggles of one breeding season. Typically, one litter (3–5 young) is born each year. Many individuals die prior to mating. Life tables from preliminary studies revealed the following patterns. There is a 50% (1/2) chance that any individual born will be female. There is a 50% (1/2) chance of an individual not surviving its first winter, and therefore not reproducing at all. There is a 17% (1/6) chance that a *surviving* individual will have three offspring, a 17% (1/6) chance that a surviving individual will have four offspring, and a 17% (1/6) chance that a surviving individual will have five offspring. If this information is correct, the population will replace itself each generation, and maintain stable population numbers over time and a stable age distribution (see chapter 6). However, recent field reports suggest that most of the populations have experienced extreme fluctuations in numbers, and many have in fact gone extinct.

Secretary of the Interior A. R. Templeton is considering listing this species as endangered. He has requested your expertise as a theoretical population geneticist: he wants you to run simulations examining how population size affects genetic diversity and extinction rates. Obviously, your simulations will be based on the known biology of the species. These simulations can also contribute to designing a conservation program for this particular species.

Simulation Rules

This simulation requires the use of one six-sided playing die and a coin. Conveniently, because the life history traits of the species fall into probabilities of approximate multiples of 1/6, we can simulate them using a six-sided die. If there is at least one male present, all females can reproduce. If any population has no males, or no females, the population goes extinct. Because you are keeping track of maternal lineage by following the mtDNA, you do not have to roll for the reproductive output of males. But you must keep track of the presence of males in each generation to record population size, and to be sure that females in the population can indeed reproduce. The rules, applied to female adults only, for the rolls of the die are:

- If you roll 1, 2, or 6, that individual dies. Cross that individual off the list.
- If you roll 3, that individual has three offspring. Now flip a coin three times to determine the sex of the three offspring: heads=female, tails=male. In column "Generation two" of table 11.1 enter the three offspring, using the mother's corresponding mtDNA label and the sex of the offspring (i.e., if the mother has mtDNA label B, then one possible outcome based on the coin flips for the three offspring she produces is: B-female, B-female, and B-male.
- If you roll 4, that individual has four offspring. Now flip a coin four times to determine the sex of the four offspring: heads=female, tails=male. Under "Generation two" enter the four offspring, using the mother's corresponding mtDNA label and the sex of the offspring.
- If you roll 5, that individual has five offspring. Now flip a coin five times to determine the sex of the five offspring: heads=female, tails=male. Under "Generation two" enter the five offspring using the mother's corresponding mtDNA label and the sex of the offspring.

A simulated group is shown in table 11.1 through three generations.

TABLE 11.1.
Sample simulation results.

Gen 1	Your die says	Coin toss	Gen 2	Your die says	Coin toss	Gen3	Etc.	Etc.	Etc.
A-f	2 (death)		—						
B-f	3 (3 pup)	Heads	B-f	2 (death)		—			
C-f	4 (4 pup)	Heads	B-f	2 (death)		—			
A-m	(don't roll)	Tails	B-m	(don't roll)		—			
B-m	(don't roll)	Tails	C-m	(don't roll)		—			
C-m	(don't roll)	Tails	C-m	(don't roll)		—			
		Heads	C-f	5 (5 pups)	Heads	C-f			
		Heads	C-f	6 (death)	Tails	C-m			
					Heads	C-f			
					Heads	C-f			
					Heads	C-f			

Small Mojave Population

The Mojave population currently consists of only 6 individuals (A-female, B-female, C-female, A-male, B-male, C-male). These individuals are listed on the Mojave results (table 11.2) in the column "Generation one."

To generate simulation data on survival and fecundity, you must first roll the die for female A. Strike female A out, or record its offspring in "Generation two" of the Row A-f (for data from female-A), according to the rules above. Refer back to table 11.1 to be sure you see how this process works. Repeat this process for female B. Enter her results in the column labeled "Generation two" in the row B-f (female B). Repeat this process for the third female in generation one and enter the results in "Generation two" along with the others. Remember that presence of males in the population is necessary, but beyond that they don't count, so don't roll the dice for them or you would be counting their offspring twice.

Now repeat this process for all the individuals of generation two, reporting the results in a new column: Generation three. Continue this process for 10 generations or until all individuals die or until there are all males or all females in the population, whichever happens first. *You need not proceed past 10 generations.* Note: When a letter is listed twice or more in a column, you must roll separately for each individual with that phenotype.

(N.B. If you don't want to roll dice, you could use the RANDBETWEEN function in Microsoft Excel. If you type "=RANDBETWEEN(1,6)" into an Excel cell and then copy this function into many other cells, they will give you a series of die rolls between 1 and 6. You could also use IF/THEN statements and a RANDBETWEEN(1,2) to assign sex. You are welcome to develop your own Excel version of the simulation, but we have not provided one because this exercise is attempting to make the simulation transparent, rather than the outcome of a black box. Of course if you make the box yourself, it won't be a mystery to you.)

TABLE 11.2.
Simulation results for the small Mojave population.

	Gen 1	Gen 2	Gen 3	Gen 4	Gen 5	Gen 6	Gen 7	Gen 8	Gen 9	Gen 10
A-f										
A-m										
B-f										
B-m										
C-f										
C-m										

Large Bakersfield Population

Now repeat this entire process in table 11.3 for the Bakersfield population of 18 individuals (9 males and 9 females), with 3 different phenotypes A, B, and C. Thus you have: A-female, A-female, A-female, B-female, B-female, B-female, C-female, C-female, C-female, A-male, A-male, B-male, B-male, B-male, C-male, C-male, C-male. *Do this for 10 generations.* Add rows to the table if the population grows too large for the given number of rows.

Questions to Work on Individually Outside of Class

1. Make two graphs. First, plot generation (*x*-axis) versus population size; second, plot generation versus allelic diversity (the number of alleles still present in the population). Each graph should have the results from both populations on it, so you can compare them easily. Use one line for Mojave and one line for Bakersfield on each graph.

   (a) Generation vs. population size. Label the *x*-axis (horizontal axis) "Generation" and number from 1 to 10. Label the *y*-axis (vertical axis) "Population size" and number appropriately. Be sure to include both sexes in these results.
   (b) Generation vs. diversity. Label the *x*-axis "Generation" and number from 1 to 10. Label the *y*-axis "Diversity" and number it from 1 to 3. A "3" means that all phenotypes are present. A "2" means that only two phenotypes are present (any two). A "1," of course, means that only one phenotype is present in the population. Be sure to include both sexes in the results.

2. Did the Mojave population go extinct? If so, in which generation?
3. Did the Bakersfield population go extinct? If so, in which generation?
4. Which population did you expect to go extinct first? Why did you expect this? Did your data support this?

TABLE 11.3.
Simulation results for the large Bakersfield population.

	Gen 1	Gen 2	Gen 3	Gen 4	Gen 5	Gen 6	Gen 7	Gen 8	Gen 9	Gen 10
A-f										
A-f										
A-f										
A-m										
A-m										
A-m										
B-f										
B-f										
B-f										
B-m										
B-m										
B-m										
C-f										
C-f										
C-f										
C-m										
C-m										
C-m										

5. Which population did you expect to lose diversity more rapidly? Why did you expect this? Does your graph show this?

6. To draw broader conclusions from simulations like this, what else would you need?

Small-Group/In Class Exercise

Pooling and Interpreting Simulation Results

Secretary Templeton has asked for a final report based on your simulations. He is interested in the effects of small population size on both the likelihood of extinction and the loss of genetic diversity. The Secretary understands that a single simulation run is meaningless, so he asked everyone in your office to run the simulations. Now your team (everyone in the class) can pool your data for populations of 6 and 18 woggles. The Secretary also assigned a summer intern over in the fisheries department to write a computer algorithm to run similar simulations for populations of 2, 10, 14, and 22 woggles. The intern ran the simulation 100 times for each of these four different population sizes, and reported the data in table 11.4. Add the collective data from your team to this table. You have each already graphed the relationship between population size and generation number for both the Mojave and Bakersfield populations. You each also noted if and when these populations went extinct within the ten generations they were followed. Now you need to calculate the "mean time to extinction" for all your replicates of the simulation for both the Mojave and the Bakersfield populations. To calculate the mean, sum all of the extinction times and divide by the number of simulations. *(If a population did not go extinct by the tenth generation simply average it in as the number 10. This is also how the computer simulation dealt with populations that did not go extinct in ten generations. However, you should consider how this may bias the result.)* You also need to calculate the standard deviation for the average time to extinction. (See also chapter 14.) You can do this on your calculator or use the equation:

$$\text{Std. dev.} = \sqrt{\frac{\sum (x_i - \text{avg})^2}{(n - 1)}} . \tag{11.1}$$

(Square the difference between each time to extinction and the average time to extinction, sum all those squared values, divide that by one less than the number of simulations, and take the square root of that quotient.)

Questions 7–10 are about small population size and extinction.

7. Graph the relationship between initial population size (on the *x*-axis) and average time to extinction (on the *y*-axis). Include the data for the two points you pooled as a group along with the data from table 11.4.

TABLE 11.4.
Time until extinction for woggle populations of various sizes.

Initial size of population	2	6	10	14	18	22
Replicates of the simulation	100	*	100	100	*	100
Average time to extinction	2.97		7.14	7.63		8.68
Standard deviation	3.03		3.15	3.04		2.21

*The number of replicates for populations starting with 6 and 18 individuals will depend on how many classmates have pooled their simulation results.

(Note: the number of replicates for populations starting with 6 and 18 individuals will depend on how many classmates have pooled their simulation results.)

8. Is there a linear relationship between population size and average time to extinction? If not, what is the relationship? What is the biological significance of this relationship?

9. If you were concerned about the impact of drift on extinction rate, but you had limited resources to design a woggle reserve, how would you pick the ideal size of the initial woggle population? What size would you select? Base your argument on the graph you just plotted.

10. If only six people repeated the simulation for the population of 22 woggles, would you expect their pooled data to have approximately the same variance or standard deviation as presented in the table below? Why or why not? Why are multiple runs of a simulation necessary?

Questions 11–16 are about small population size and biodiversity.

11. Previously you plotted the relationship between "generation time" and the "number of different alleles present" for both the Mojave and the Bakersfield populations. You now need to plot the relationship between "generation time" and the "average number of different alleles present." To calculate the average number of different alleles present, record the generation number in which the population first dropped to two alleles for each member of the group, and average these values. Do the same when it drops to one allele and then to zero alleles.

12. Did every population lose the alleles in the same order? Why or why not?

13. If several isolated populations initially start with the same gene frequencies (as was the case for the Mojave and Bakersfield populations on which you ran simulations), what would you expect the impact of drift to be on the genetic variation between populations? (Remember that the gene frequency is simply the proportion of a particular gene in the population at a particular time.)

14. What if the populations examined were not completely isolated, but in fact some migration occurred between populations? What would you expect the impact of migration to be on the genetic variation between populations?

15. What other data about the species would help you develop more reliable simulations? Be specific in describing what information you need, and explain how you would use it. How would you make the simulation model more realistic? What are the drawbacks of the current model, which focuses only on genes carried by females?

16. Make a case to the Secretary for commissioning further research on this species. He has said that there is money in the budget for simulation studies. Make the case for purchasing a new computer to run the simulations. What could you do with computer simulations that you cannot realistically do with dice?

Your group has the full class period to draft the report to the Secretary. Organize your group report around the numbered questions above. Hand in one neat legible copy of the report containing answers to the questions posed above along with the graphs for questions 7 and 11. Put the names of all group members on this document. Each group member should also hand in the results of the simulations performed at home along with graphs and answers to questions 1–8 of this exercise. Make sure your name is on the front of those pages.

Reference

Gotelli, N. 1998. *A Primer of Ecology.* Sunderland, Mass.: Sinauer Associates.

12 Conservation of Small Populations: Effective Population Sizes, Inbreeding, and the 50/500 Rule
Luke J. Harmon and Stanton Braude

Introduction and Background

Population size is extremely important in evaluating conservation priorities for a species. Small populations are at risk of going extinct because of demographic stochasticity and genetic drift. In this exercise, you will learn about three of the meanings of "effective population size" and how to estimate two of them. You will then learn how to apply these techniques to specific conservation situations, using the concepts of inbreeding, the minimum viable population size, and the 50/500 rule.

Effective Population Size

Population size has a major impact on the dynamics of a population. For example, in chapter 11 you used simulations to see that genetic drift reduces allelic diversity much faster in small populations of woggles than in large ones. Population size also influences the chances of extinction through demographic stochasticity, the random change in population size over time due to random variation in individual survival and reproductive success. Such events have a proportionally large effect in small populations. For example, in a population of 10 individuals, one accidental death would reduce the population size by 10%. In contrast, if the population were made up of 1000 individuals, one accidental death would reduce the population size by only 0.1%. Thus, small populations are much more likely to go extinct due to demographic stochasticity than are large populations.

Effective population size (N_e) helps us quantify how a particular population will be affected by drift or inbreeding. Effective size takes into account not only the current census size of a population, but also the history of the population. Effective population size is the size of an "ideal population" of organisms (ideal refers to a hypothetical population in the Hardy Weinberg sense with a constant population size, equal sex ratio, and no immigration, emigration, mutation, or selection) that would experience the effects of drift or inbreeding to the same degree as the population we are studying. For example, if our actual population of 50 animals experiences the effects of drift at the same rate as an ideal population of 20 animals, the population has a drift effective size of 20.

There is no such thing as "*the* effective size" of a population. Different effective population sizes help us estimate the impact of different forces. The effective size you estimate will depend on the scientific question you are trying to address (box 12.1). Estimating the appropriate effective population size is crucial in conservation biology; in most (but not all)

Box 12.1 Different Ways to Measure Effective Population Size

There are a variety of population effective sizes that have different mathematical and biological meanings. The terms are sometimes confused or misunderstood as synonymous. Such confusion can have serious implications for understanding and managing populations of endangered or threatened species, as we see below.

Inbreeding effective size, N_{ef}, refers to the size of an ideal population that would allow the same accumulation of pedigree inbreeding as the actual population of interest. Pedigree inbreeding occurs when an offspring inherits two copies of a gene from its parents which are identical by descent—that is, they are both directly descended from a single allele present in one of the founders of that population (perhaps the parents are cousins and each inherited the particular allele from the same grandfather). N_{ef} is the measure of effective population size that emphasizes the effect that small population size has on the chances of relatives mating with each other. Such matings lead to a loss of heterozygosity in the population. Thus, this effective size gives you an indication of the likely loss of heterozygosity across all alleles in your population.

Calculation of N_{ef} ideally requires pedigree data. However, you can estimate the inbreeding effective population size (N_{ef}) by calculating the *harmonic mean* of the population size over time from the founding generation to the penultimate generation. The symbol t represents the number of generations for which we have population size data. $N(0)$ is the size of the founding population, $N(1)$ is the size of the population after one generation etc. and, $N(t-1)$ is the size of the population one generation ago.

$$(a)\ N_{ef} = \frac{t}{\dfrac{1}{N(0)} + \dfrac{1}{N(1)} + \cdots + \dfrac{1}{N(t-1)}}$$

Variance effective size, N_{ev}, refers to the size of an ideal population that would accumulate the same amount of variance in allele frequencies as the population of interest; thus, this effective population size indicates how rapidly allele frequencies are likely to change. This is important because it also affects how rapidly isolated populations diverge from one another under genetic drift. Again, the symbol t represents the number of generations for which we have population size data. $N(1)$ is the size of the population after one generation, etc., and $N(t)$ is the size of the current population.

$$(b)\ N_{ev} = \frac{t}{\dfrac{1}{N(1)} + \dfrac{1}{N(2)} + \cdots + \dfrac{1}{N(t)}}$$

In addition, the following correction can be used at each generation if operational sex ratios are not 1:1. This corrected population size reflects the increased effects of both inbreeding and drift when the sexes are not contributing equally to the allele pool.

$$(c)\ N_s = \frac{4 N_m N_f}{N_m + N_f}$$

(continued on following page)

Notice the difference between formula (a) and formula (b). While the *inbreeding effective size* is more sensitive to the *number of original founders* [$N(0)$], the *variance effective size* is more sensitive to the *number of offspring in the current generation* [$N(t)$]. This is because, as stated above, N_{ef} focuses on the loss of heterozygosity due to pedigree inbreeding in the population; with a small initial founding population, close relatives are likely to mate with each other. In contrast, N_{ev} gives an indication of the increase in variance of allele frequencies between subpopulations due to drift, and depends on the number of offspring produced by those founders and by each subsequent generation, up to the present-day $N(t)$.

These differences can lead to large discrepancies between these two different effective population sizes in real populations. For example, *increasing populations* generally have a larger N_{ev} than N_{ef}, while *declining populations* will generally have larger N_{ef} than N_{ev}. Hence, a population coming through a bottleneck may have a low inbreeding effective size, but it can have a larger variance effective size if the population bounces back rapidly (as is the case with the Southern white rhinoceros population, which you will work on in class).

There are other measures of effective population size that focus on different population genetic parameters. For example, *eigenvalue effective size, $N_{e\lambda}$*, focuses on the rate at which unique alleles are lost from a population. For this exercise, we will examine only the two effective population size estimates discussed above.

cases, effective population size will be smaller than the actual number of organisms in the population. Think for a moment about why this is so. A conservative rule of thumb used by some biologists is that N_e is usually about one-fifth of the total population size (Mace and Lande, 1991). Using such a rough estimate is risky because N_e can be *larger* than the census size of the population, depending on the history of the population and the particular N_e under consideration.

Demographic stochasticity, genetic drift, and environmental variation can interact to doom a small population to extinction. This is called an extinction vortex, and it is due to a positive feedback loop (figure 12.1): the negative consequences of lower effective population size make the population smaller, causing stronger negative effects, leading to an even smaller population size (Gilpin and Soule, 1986). For example, a random environmental change might lower population size, leading to a higher chance of population reduction due to demographic stochasticity. This could lower inbreeding effective population size even more, leading to severe inbreeding depression and reduced fertility. This further reduces the population size. Chains of events such as these mean that the extinction probability for a small population can be extremely high. For example, Pimm et al. (1988) showed that the extinction risk for birds on small islands off the coast of Britain rises with decreasing numbers of nesting pairs. Conservation biologists realize that an extinction vortex can begin when humans cause major reductions in the population size of a species.

Calculating Effective Population Sizes

Consider the data in table 12.1 for a population of Eastern fence lizards, *Sceloporus undulatus*, at Tyson Research Center in eastern Missouri: You can estimate the inbreeding effective

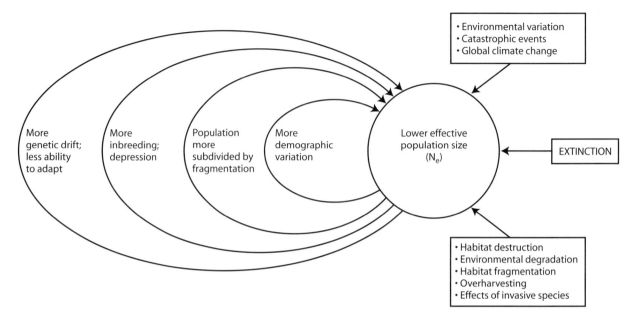

FIGURE 12.1. Extinction vortex. Population size decreases in a positive feedback loop, eventually resulting in the extinction of the population (from Primack, 2000).

size for this population using formula (a) from box 12.1. In this case, $t=4$ generations; we calculate the *inbreeding effective population size* as:

$$N_{ef} \approx \frac{t}{\dfrac{1}{N(0)} + \dfrac{1}{N(1)} + \dfrac{1}{N(2)} + \dfrac{1}{N(3)}}$$

$$= \frac{4}{\dfrac{1}{140} + \dfrac{1}{250} + \dfrac{1}{110} + \dfrac{1}{26}} = \frac{4}{0.058} = 68.9 \approx 69$$

(12.1)

Try this calculation yourself—it can get confusing taking all of these reciprocals!

This estimated effective population size (69) means that, in terms of genetic inbreeding, this population (with a mean census size of 148 lizards over 5 years) will accumulate the

TABLE 12.1.
Sceloporus population at Tyson from 1996 to 2000.

Year	Population count
1996	140
1997	250
1998	110
1999	26
2000	180

effects of inbreeding at the same rate as a population that had a constant size of only 69 individuals.

In contrast, the *variance effective size*, estimated with formula (b) in box 12.1, is

$$N_{ev} \approx \frac{t}{\dfrac{1}{N(1)} + \dfrac{1}{N(2)} + \dfrac{1}{N(3)} + \dfrac{1}{N(4)}}$$

$$= \frac{4}{\dfrac{1}{250} + \dfrac{1}{110} + \dfrac{1}{26} + \dfrac{1}{180}} = \frac{4}{0.05655} = 70.73 \approx 71.$$

(12.2)

This effective population size (71) means that, in terms of genetic drift, this population (with a mean size of 148 over 5 years) will accumulate the effects of drift at the same rate as a population that had a constant size of 71 individuals. Note that, in this example, the mean size of the population over time reflects neither how it will accumulate inbreeding nor how it will experience drift. Although the difference between variance effective size and inbreeding effective size appears small in this example, the point is that these are not two ways of estimating *the* effective size; these are two different effective population sizes. In the examples you work with in this exercise, you will see how one population can have very different effective sizes that inform us about very distinct risks for the population.

Effective Population Size, Inbreeding, and Extinction

Just as there are a number of different meanings of effective population size, there are a number of different meanings of inbreeding. Because breeding with close relatives typically reduces the pool of genes contributing to the next generation, one measure of inbreeding is F, inbreeding as a measure of drift. In addition to increasing the impact of drift, inbreeding can increase the proportion of deleterious homozygous gene combinations in a population, which leads to lower survival of young and thus lower reproductive output; this is called inbreeding depression. To begin to evaluate the potential impacts of small population size on inbreeding, we can use the following estimation (from Soule, 1980):

$$\Delta F = 1 - \left[1 - \frac{1}{2N_{ef}} \right] t.$$

(12.3)

In this equation, ΔF is the increase in the inbreeding coefficient over time; N_{ef} is the inbreeding effective population size; and t is the number of generations. We can use this equation to predict how much the inbreeding coefficient in a small population will increase over time. If ΔF is 0.6 or higher, the fecundity of individuals in the population may be reduced, and the population could be at higher risk for extinction. For example, we can calculate the increase in the inbreeding coefficient after 100 generations for the population described above. For the fence lizards, N_{ef} was 69. This means that $\Delta F = 1 - (1 - 1/138)^{100} = 0.52$. We would conclude that after 100 generations the fence lizards are experiencing some inbreeding; however, because $\Delta F < 0.60$ it may not threaten the population. Of course, the inbreeding coefficient will continue to increase over time, possibly to dangerous levels, if the population remains small and isolated. Note that the estimate of ΔF ignores gene flow—and even moderate gene flow can greatly reduce the effect of genetic drift, slowing the rate of increase in the inbreeding coefficient over time. So even moderate gene flow can help maintain genetic diversity. This is one reason why many conservation biologists advocate the maintenance of corridors connecting small, isolated populations of a species.

Minimum Viable Populations and the 50/500 Rule

You know that demographic stochasticity and genetic drift can negatively affect small populations. Demographic stochasticity leads to the random extinction of small populations, while genetic drift can cause a reduction of genetic diversity within a population. These factors can interact in an extinction vortex (figure 12.1 discussed above) that can eventually lead to the extinction of a population.

To decide when these factors might be important for a population of an endangered species, Shaffer (1981) proposed the concept of the minimum viable population (MVP). He defined the MVP as the smallest isolated population (of a given species in a given habitat) having a 99% chance of remaining in existence for 1,000 years, despite the foreseeable effects of demographic stochasticity, genetic drift, environmental stochasticity (random changes in the environment), and natural catastrophes (Shaffer, 1981). Shaffer chose the percentage and time scale to represent what most scientists consider a good chance for survival of a species. Quantitative objectives like the MVP provide specific guidelines for gauging the success of conservation programs (Foose et al. 1995). Populations smaller than the MVP are considered to be at significant risk of entering into the extinction vortex and becoming extinct, so a conservation program can be considered successful only if it raises the effective population size above the MVP.

A related concept is the 50/500 rule, proposed by Franklin (1980). The "50" part of the 50/500 rule states that populations with an inbreeding effective population size (N_{ef}) under 50 are at immediate risk of extinction. This is because, in such small populations, inbreeding and demographic stochasticity can quickly push the population into an extinction vortex. The "500" part of the rule means that populations with a variance effective size (N_{ev}) of less than 500 are at long-term risk of extinction. In these populations, genetic drift may be a strong force, leading to eventual loss of genetic variation (Franklin, 1980). After variation is lost, the population will no longer be able to respond to environmental changes, and may be reduced in size or go extinct if any such changes occur. Even when properly understood, Franklin's rule is quite controversial. Some authors question its generality, and others suggest that the numbers are too small. For example, Lande (1995) suggested that any population with N_{ev} less than 5,000 will be subject to strong genetic drift, which will deplete the genetic variation in a population and cause long-term extinction risk. Recent experimental tests of the rule in captive housefly populations also suggest that populations must be higher than implied by the 50/500 rule in order to survive and maintain genetic diversity (Reed and Bryant, 2000). As a conservation biologist, you may not use the 50/500 rule, but it is essential that you understand where the 50 and 500 come from.

Homework for this exercise takes approximately 45 minutes.

Case Studies and Data

European Adders

The European adder (*Vipera berus*) is a small, venomous snake (figure 12.2) distributed throughout Europe (Arnold and Burton, 1978). This snake occupies a wide geographic range, but within that range snakes are often found in small, isolated populations separated from each other by hundreds to thousands of meters. Under natural conditions, gene flow between these subpopulations is high, because the male snakes disperse widely when

FIGURE 12.2. European adder *Vipera berus*.

searching for females during the spring (Madsen et al., 1993). However, in certain regions humans have disturbed adder habitat through agriculture and urban encroachment. Now we see extremely isolated small populations of adders, surrounded by large areas of unsuitable habitat (Madsen et al., 1996). Consider the isolated population of adders at Smygehuk, on the south coast of Sweden. These adders are separated from the nearest population by 20 km of farmland, which is unsuitable habitat for adders (Madsen et al., 1996). Use the data in tables 12.2 and 12.3 (Madsen et al., 1996) to answer the following questions.

Questions to Work on Individually Outside of Class

1. Table 12.2 gives data on the population size of the adders for each year from 1984 to 1990.

   (a) Plot the total number of adult adders (y) over time (x).
   (b) Use these data to calculate the inbreeding effective size and the variance effective size of this population of adders.
   (c) Explain why the inbreeding effective size and the variance effective size of this population differ.
   (d) Recalculate the variance effective size of this population with the new information on the sex ratios of snakes in the population (table 12.3).
   (e) Dr. Fern Skipe has argued that there is no such thing as the "real" effective population size. Do you agree with this statement? Why or why not? Use the results above in your argument.
   (f) Based on these data and your calculations, make a recommendation to the Swedish government concerning this population of vipers.

TABLE 12.2.
Total number of adult adders at Smygehuk

Year	Total number of adults
1984	138
1985	40
1986	34
1987	42
1988	37
1989	41
1990	34

TABLE 12.3.
Adult male and female adders in each year

Year	No. of adult females	No. of adult males	Corrected population size
1984	98	40	
1985	29	11	
1986	24	10	
1987	32	10	
1988	27	10	
1989	29	12	
1990	27	7	

Small-Group/In-Class Exercise

Please bring graph paper and a calculator to class to complete the next part of this exercise. First, read the following background information on African rhino conservation.

African Rhino Conservation

Background The rhinoceros is an example of a "charismatic megavertebrate" that has played a central role in promoting worldwide conservation efforts. The five extant species of rhino are the last representatives of a large group of species that reached a peak in diversity between 25 and 5 million years ago (Estes, 1991). Two of the five extant species occur in Africa: the white rhinoceros (*Ceratotherium simum*) and the black rhinoceros (*Diceros bicornis*). Despite their names, both species are a dull gray color, and can be distinguished by the shape of their mouthparts (figure 12.3). The black rhino has a hook-shaped triangular upper lip that allows it to obtain its food by browsing leguminous herbs and shrubs. The white rhino, on the other hand, has a very wide, square mouth, and is specialized in grazing areas of dense grasses. All rhinos have poor eyesight and relatively small brains, but ex-

FIGURE 12.3. Black (left) and white (right) rhinoceri.

tremely sensitive hearing and smell (Estes, 1991). Both African species of rhino show geographic variation. The black rhino has been divided into four subspecies, western (*Diceros bicornis longipes*), eastern (*D. b. michaeli*), southwestern (*D. b. bicornis*), and south central (*D. b. minor*). The western subspecies is the rarest and most isolated, with only a few individuals living in western Africa. The white rhino has been divided into two subspecies, the northern (*Ceratotherium simum cottoni*) and southern (*C. s. simum*). These two subspecies occupy separate ranges and are more distinct, both morphologically and genetically, than the subspecies of black rhinos (Emslie and Brooks, 1999).

Both species of rhino have undergone major reductions in their ranges in the past several hundred years. Early colonial explorers reported that black rhinos were widespread in distribution and fairly common, while white rhinos were more restricted in range. After European colonization, southern white rhinos were very quickly reduced to near-extinction, reaching a low of just 20 individuals in 1895. Since then, numbers of the southern white rhino have steadily increased; there are over 8,000 alive today. The northern white rhino, on the other hand, has shown a dramatic decrease in recent years, declining from over 2,000 in 1960 to only 25 individuals in 1998. Numbers of black rhinos have also declined since colonial times. Declines were especially severe between 1970 and 1992, when black rhinos declined 96%. The species has recently shown some potential for recovery (Emslie and Brooks, 1999). The main reason for the decline of all rhinos is hunting by humans. European colonists killed hundreds of thousands of rhinos during the nineteenth century. More recently, rhinos have been killed by poachers supplying markets in Asia and the Middle East with rhinoceros horns. In Asia, rhino horns are used in traditional Chinese medicine, whose practitioners believe that the horns lower fevers, increase male potency, and can cure a host of diseases. In the Middle East, they are used as handles for ornamental daggers called jambiyas (Emslie and Brooks, 1999). Although rhino horns have been used for these purposes for hundreds of years, recent increases in demand put serious pressure on wild rhinos, and poaching for horns is the major threat to African rhino populations today (Emslie and Brooks, 1999).

***Black Rhinoceros* Diceros bicornis** Black rhinos were formerly the most widespread and abundant species of rhino (Estes, 1991), but are now listed in the IUCN Red Book as critically endangered. Direct counts of black rhinos have shown declines of over 80% in

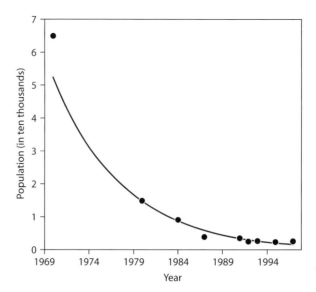

FIGURE 12.4. The black rhinoceros has been in severe decline in recent years and is entering a population bottleneck (data from Emslie and Brooks, 1999).

the last 50 years (Emslie and Brooks, 1999). The total population size of black rhinos in 1970 was estimated at 65,000. Historically, black rhinos occupied a large range throughout Africa. Today, black rhino populations are fragmented, and the species is rapidly declining in numbers (figure 12.4). The species has been divided into four subspecies, each occupying a separate geographic range: the western, eastern, southwestern, and south-central black rhinos. The ranges of each subspecies have different climates and habitats. They can sometimes be distinguished by characters such as skin texture and horn length and shape. There may also be genetic and behavioral differences across these subspecies. The western black rhinoceros is the most range-restricted and endangered of all the subspecies of black rhinos: the entire subspecies is represented by only a few scattered individuals in Cameroon. This subspecies, separated from the rest of the black rhinos by hundreds of miles, may represent a genetically distinct lineage, but is threatened with extinction in the immediate future. The largest population of black rhinos is in Kenya, while the main population of southwestern black rhinos is in Namibia. The south-central black rhino is the most common subspecies, with large numbers in South Africa and Zimbabwe, and smaller populations in southern Tanzania and Mozambique. The goal of your group is to establish a conservation plan for black rhinos as a whole. You need to decide how much money needs to be allocated to each of the four subspecies, and which populations will have the highest conservation priority.

***White Rhinoceros* Ceratotherium simum** White rhinos have always been less common and more limited in distribution than black rhinos (Emslie and Brooks, 1999). This is probably a function of their specialized grazing diet. This feeding strategy makes the white rhino unique, as it is quite different from that of black rhinos and, in fact, from all other rhinos in the world (Estes, 1991). The white rhino is one of the largest purely grazing herbivores that has ever lived. Southern white rhinos are separated from northern white rhinos by 2,000 km, and no white rhino has ever been recorded in the intervening area. These two subspecies are genetically distinct, with more genetic variation between these two subspecies than among the four subspecies of black rhino (Smith et al., 1995).

Southern White Rhinoceros **Ceratotherium simum simum** Southern white rhinos were on the brink of extinction in 1895, when overhunting had reduced them to just 20 individuals in one population in South Africa. Their numbers have since recovered substantially (figure 12.5). The recovery of the southern white rhino is one of the major success stories in modern conservation biology. Before their decline in the nineteenth century, their range included much of southern Africa. After initial recovery of the source population, many translocations were carried out; these have successfully reintroduced rhinos into areas where they had been wiped out. Southern white rhinos are now the most numerous subspecies of rhino; populations can be found in South Africa, Botswana, Namibia, Swaziland, and Zimbabwe. They have also been introduced into Kenya, Ivory Coast, and Zambia, all outside their native range. The rhinos have functioned as a major source of funding in South Africa, where national parks have sold excess rhinos to private game parks for as much as U.S.$25,000 apiece.

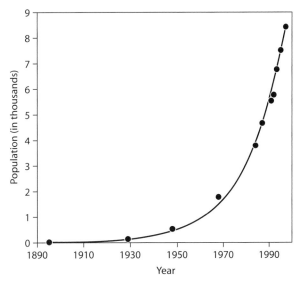

FIGURE 12.5. The southern white rhinoceros has recovered from near extinction at the turn of the last century and the current growing population is descended from a bottleneck population of only 20 animals (data from Emslie and Brooks, 1999).

Northern White Rhinoceros **Ceratotherium simum cottoni** Northern white rhinos have shown a striking decline in recent years, and are now perilously close to extinction. Table 12.4 gives data on northern white rhinoceros populations by country from 1960 to 1998 (Emslie and Brooks, 1999). Although the range of northern white rhinos once included parts of Uganda, Chad, Sudan, the Central African Republic, and the Democratic Republic of Congo, they are now restricted to a small area in the northeast of the Democratic Republic of Congo (Emslie and Brooks, 1999). This population numbered only 25 individuals in 1998, and DRC has been suffering from tremendous political instability over the past 20 years. Conservation biologists have not lost hope of preserving the subspecies,

TABLE 12.4.
Northern white rhinos by country, 1960–1998.

	1960	1971	1976	1981	1983	1984	1991	1995	1998
Central African Republic	Few	Few	Few	Few	Few	0?	—	—	—
Chad	Few	Few	?	?	0?	0?	—	—	—
Democratic Republic of the Congo	1,150	250	490	<50	13–20	15	30	31	25
Sudan	1,000	400	?	<300	<50	0?	0?	0?	0?
Uganda	80	Few	Few	Few	2–4	0?	—	—	—
Total	2,230	650	500+	<350	<70	15	30	31	25

however, and cite the recovery of the southern white rhino, which has grown from a total population of around 20 individuals in 1895 to over 8000 today. The northern white rhinos currently surviving in the Democratic Republic of Congo represent the last survivors of a unique lineage of rhinos; their extinction would be a great and irreversible tragedy.

Your Job: Help Create Species Survival Plans for African Rhinos (Questions 2–6)

Species Survival Plans (SSPs) coordinate the management of rare and endangered species to maintain healthy breeding populations, retain genetic variation, and minimize "inbreeding." SSPs often have the conflicting goals of preserving species in a captive environment while at the same time minimizing evolutionary change in the species and minimizing loss of genetic diversity from inbreeding or drift (Templeton, 1991). These can be significant forces affecting wild (in situ) and captive populations that are entering or emerging from population bottlenecks. Your job will be to use real data to help the IUCN Rhino-Rescue Team generate an SSP for the two species of African rhinos. After you answer questions 2–5, the entire class will meet as a committee of the whole to allocate funds for the conservation of African rhinos.

2. Your instructor will assign you to one of the three rhino species or subspecies on which we have data. First, you need to assess the current genetic situation for black rhinos, or southern or northern white rhinos. This assessment should include the inbreeding and variance effective sizes for the wild populations. You should be able to project accumulation of inbreeding in wild populations if they are maintained at current levels. (Assume equal sex ratios and a generation time of 8 years. For black rhinos and southern white rhinos you will need to estimate census sizes from figures 12.4 and 12.5) Your instructor will copy table 12.5 on the board and you can share your results with the other groups.

3. For your species or subspecies, discuss the long-range plan for maintaining the genetic health of the population. Address the recommendations and the theoretical framework of Franklin's 50/500 rule in your plan.

4. You should also discuss the situation for your species in the wild and decide whether you want to use the wild population to supplement the captive zoo population or vice versa.

5. Finally, your plan must include priorities for both species and for different populations within each species. The reality is that there are limited funds available for rhino conservation, and you must generate guidelines about where resources should be spent.

TABLE 12.5.
Population census and effective sizes of African rhinos.

	Census size, 1997	Inbreeding effective size (N_{ef})	Variance effective size (N_{ev})
Black rhinoceros *Diceros bicornis*	$N = 2,600$		
Southern white rhinoceros *Ceratotherium simum simum*	$N = 8,440$		
Northern white rhinoceros *Ceratotherium simum cottoni*	$N = 23$		

6. Each group will have a few minutes to describe the situation for their rhinos and propose an allocation of the $500,000 which African Rhino Rescue has raised. Once each group has made their brief presentation you will have a chance to convince each other (and your instructor) to take your recommendation.

References

Arnold, E. N., and J. A. Burton. 1978. *A Field Guide to the Reptiles and Amphibians of Britain and Europe.* London: Collins.

Foose, T. J., L. de Boer, U. S. Seal, and R. Lande. 1995. Conservation management strategies based on viable populations. Pages 273–294 in: J. D. Ballou, M. E. Gilpin, and T. J. Foose (eds.), *Population Management for Survival and Recovery: Analytical Methods and Strategies in Small Population Conservation.* New York: Columbia University Press.

Franklin, I. R. 1980. Evolutionary change in small populations. Pages 135 – 140 in: M. E. Soule and B. A. Wilcox (eds.), *Conservation Biology: An Evolutionary-Ecological Perspective.* Sunderland, Mass.: Sinauer Associates.

Gilpin, M. E., and M. E. Soule. 1986. Minimum viable populations: processes of species extinction. Pages 19–34 in M. E. Soule (ed.), *Conservation Biology: The Science of Scarcity and Diversity.* Sunderland, Mass.: Sinauer Associates.

Hendrik, P. W. 2000. *Genetics of Populations.* 2nd ed., Sudbury, Mass.: Jones and Bartlett Publishers.

Lande, R. 1995. Mutation and conservation. *Conservation Biology* 9: 782–791.

Madsen, T., R. Shine, J. Lomanmm, and T. Håkansson. 1993. Determinants of mating success in male adders, *Vipera berus. Animal Behavior* 45: 491–499.

Madsen, T., B. Stille, and R. Shine. 1996. Inbreeding depression in an isolated population of adders *Vipera berus. Biological Conservation* 75: 113–118.

Mace, G. M., and R. Lande. 1991. Assessing extinction threats: Towards a reevaluation of IUCN threatened species categories. *Conservation Biology* 5:148–157.

Pimm, S. L., H. L. Jones., and J. Diamond. 1988. On the risk of extinction. *American Naturalist* 132: 757–785.

Primack, R. B. 2000. *A Primer of Conservation Biology.* Sunderland, Mass.: Sinauer Associates.

Reed, D. H., and E. H. Bryant. 2000. Experimental tests of minimum viable population size. *Animal Conservation* 3: 7–14.

Shaffer, M. L. 1981. Minimum population sizes for species conservation. *Bioscience* 31: 131–134.

Soule, M. E. 1980. Thresholds for survival: maintaining fitness and evolutionary potential. Pages 151–169 in M. E. Soule and B. A. Wilcox (eds.) *Conservation Biology: An Evolutionary-Ecological Perspective*. Sunderland, Mass.: Sinauer Associates.

Walpole, M., M. Morgan-Davies, S. Milledge, P. Bett, and N. Leader-Williams. 2001. Population dynamics and future conservation of a free-ranging black rhinoceros (*Diceros bicornis*) population in Kenya. *Biological Conservation* 99: 237–243.

Dispersal and Metapopulation Structure

James Robertson

Introduction and Background

One of the greatest concerns facing conservation biologists today is the continued expansion of human development into natural areas. This urban and suburban sprawl occurs at a great price to the landscape and natural ecosystems. Farms, logging roads, housing developments, shopping malls, and highways now subdivide natural landscapes that were once larger and more continuous. Highly fragmented habitats often contain species that cannot move through disturbed habitat. Many of these populations have previously been very large, but are now subdivided into numerous small populations, each of which is highly susceptible to the impacts of drift (see chapter 11).

Gene flow between populations, however, can redistribute genetic variation among populations, and restore local genetic variation lost through drift. Gene flow between populations can even allow recolonization where populations have gone extinct, due to a variety of stochastic processes. What is the optimal geographic distribution of protected habitat for the design of reserves and corridors? To answer, we must understand the metapopulation dynamics of affected populations. This is why conservation biologists are particularly interested in understanding the balance between gene flow and drift in both natural and recently fragmented populations. For example, the Northwest Forest Plan, which regulates forest use in the Pacific Northwest, owes much of its success to using metapopulation and landscape-scale models (Wintle et al., 2005).

Objectives of This Exercise

As you may remember from chapter 11, the fictional woggle (*Treborus treborsonii*) exists in scattered, sometimes very small, populations in California. These populations experience extreme size fluctuations over short periods of time. You know from chapter 11 that stochastic (random) fluctuations, such as a "run" of deaths, can cause the extinction of tiny populations. But dispersal can increase small local populations and recolonize empty areas. In this exercise we will examine the effect of different dispersal rates on drift and distribution of alleles in a metapopulation. As noted in chapter 11, woggles are fictional rodents, rather like kangaroo rats. The simulations in this exercise can be completed with a six-sided die, because probabilities of events have been contrived as multiples of 1/6.

Case Study and Data

Over 30 woggle populations have been sampled; estimated population sizes have varied greatly, from 3 to 300 individuals. Because so many of the known populations are small, scholars are concerned with the influence of stochastic processes on genetic variability and local extinction. Woggles have a long evolutionary history of extremely harsh environmental conditions (e.g., heavy predation by raptors and snakes), and extreme and unpredictable (see chapter 18) fluctuations in weather conditions. These conditions have resulted in an average life span for woggles of one breeding season. Typically, one litter (3–5 young) is born each year. Many individuals die prior to mating.

Life tables from preliminary data revealed the following probabilities. There is a 50% (1/2) chance that any individual born will be female. There is a 50% chance of an individual not surviving its first winter, and therefore not reproducing at all. If a female reaches reproductive age, there is a 17% chance that she will have three offspring, a 17% chance that she will have four offspring, and a 17% chance that she will have five offspring. If this information is correct, the population will replace itself each generation, and maintain stable (constant) population numbers over time.

From your work in chapter 11, you are not surprised that recent field reports suggest most of the populations have experienced extreme fluctuations in numbers, and many have gone extinct. A very recent field study conducted by world-renowned conservation biologist Mark Trail yielded the following new information. Historically, woggles existed in large continuous habitats. Now, the habitat is fragmented; many small, partially isolated populations (demes) exist (figure 13.1). There is gene flow between demes. Occasionally a deme goes extinct, presumably due to stochastic processes—but later the area maybe recolonized by a neighboring deme. We call a system like this, where many demes are connected by recurrent gene flow, and associated with local extinction and recolonization, a metapopulation.

Simulation

Congratulations! As a result of your previous work in chapter 11, Secretary of the Interior A. R. Templeton has listed *Treborus treborsonii* as an endangered species. This is a crucial step to force immediate implementation of conservation efforts, which are critical for the survival of this species. Secretary Templeton has requested your expertise as a theoretical population geneticist. He asks you to run new simulations to examine the question: how will gene flow affect the distribution of alleles among the demes of this metapopulation? Your team will compare the balance between gene flow and drift, and will determine how various gene flow rates influence allelic distribution among demes. You will calculate the "RR statistic" (similar to the F_{ST} used in population genetics, see chapter 10) to measure allelic distribution for each generation in your metapopulation. Your conclusions will be used to design a reserve for woggles.

Rules

This simulation has been designed to mimic dispersal from deme to deme in a metapopulation. Dispersal (more specifically, gene flow) among demes redistributes alleles. This gene flow reduces genetic variation among demes, making them more similar genetically. We will consider four different dispersal rates (*d*); however, each of you will be asked to run the simulation for only one *d* value. Different groups will use different *d* values in simula-

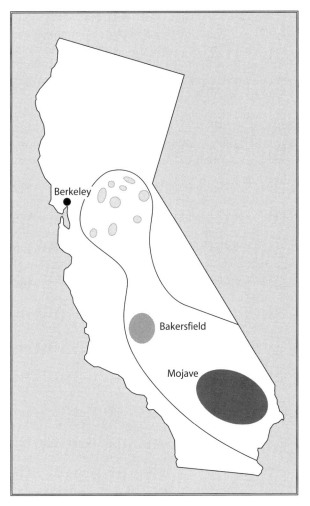

FIGURE 13.1. The distribution of the fictional California desert woggle *Treborus treborsonii*. The historic distribution is outlined and the three remaining metapopulations (Berkeley, Bakersfield, and Mojave) are shaded.

tions, and you will share your results in class. (The particular d value that you will use in this simulation will be assigned by your teacher. Each group of students will use the same *d* value but each person will run his or her own simulation.)

This simulation requires the use of at least one six-sided playing die and a coin. Conveniently, the life history traits of the species fall into probabilities of approximate multiples of 1/6 so we can simulate them with a six-sided die. So long as there is at least one male present, all females can reproduce. If any population has no males or no females, that population goes extinct. You are keeping track of maternal lineages by following mtDNA, so you do not have to roll for the reproductive output of males (this is the standard convention for demographic analysis of sexual species). You must, however, keep track of the presence of males in each generation to determine whether females have potential mates, and to calculate the genetic diversity present in each deme each generation.

This metapopulation consists of three demes. Initially, all three demes are identical. Each deme begins with the same population size, the same allele frequency, and an even sex ratio. Each deme starts with the following mitochondrial (mt) alleles: (A-f, B-f, C-f, D-f, E-f, A-m, B-m, C-m, D-m, E-m), where "f" is a female with the associated allele and "m" is a male with the associated allele. This means there are 10 woggles, 5 females and 5 males, in each deme (figure 13.2).

Dispersal Dispersal rates can range from 0 to 0.17 (3/18). We will simulate four different levels of dispersal among demes (d=0, 1/18, 2/18, or 3/18), but each person will run the simulation for only one level of dispersal.

- If you are assigned d=0, then there is no dispersal among demes. For your entire simulation, you should follow only the reproduction rules.
- If you are assigned d=1/18, there is dispersal among demes; you must apply dispersal rules to each individual in each generation before determining reproduction. A die must be rolled twice to determine whether an individual will disperse (or two dice may be rolled at the same time). "Doubles" means that you rolled the same number twice in a row (or if rolling two dice, both dice display the same number). As noted in chapter 11, you may prefer to use the random number generator in Excel rather than rolling dice. If you roll double 1s or double 2s then that individual will disperse. Now flip a coin to determine where this individual will migrate. If you flip heads, that individual moves to the adjacent deme counterclockwise. If you flip tails, that individual moves to the adjacent deme clockwise.
- If you are assigned d=2/18 there is dispersal among demes; you must apply dispersal rules to each individual in each generation before determining reproduction. A die must be rolled twice to determine whether an individual will disperse (or two dice may be rolled at the same time). "Doubles" means that you rolled the same number twice in a row (or if rolling two dice, both dice display the same number). If you roll double 1s, double 2s, double 3s, or double 4s, that individual will disperse.

    Now flip a coin to determine where this individual will migrate. If you flip heads, that individual moves to the adjacent deme counterclockwise. If you flip tails, that individual moves to the adjacent deme clockwise.

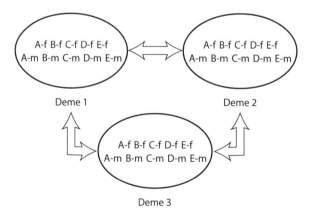

FIGURE 13.2. This simulation examines the effect of migration on loss of alleles and extinction of demes.

- If you are assigned $d=3/18$ there is dispersal among demes; you must apply dispersal rules to each individual in each generation before determining reproduction. A die must be rolled twice to determine whether an individual will disperse (or two dice may be rolled at the same time). "Doubles" means that you rolled the same number twice in a row (or if rolling two dice, both dice display the same number). If you roll double 1s, double 2s, double 3s, double 4s, double 5s, or double 6s that individual will disperse.

  Now flip a coin to determine where this individual will migrate. If you flip heads, that individual moves to the adjacent deme counterclockwise. If you flip tails, that individual moves to the adjacent deme clockwise.

Note: You must complete the dispersal simulation for all individuals of each deme (demes 1, 2, and 3), before applying the reproduction rules. All movement among demes happens before reproduction in each generation. Each individual can only disperse once each generation.

Reproduction

The rules, applied to female adults *only*, for interpreting the rolls of the die are as follows:

- If you roll 1, 2, or 6, that female does not reproduce.
- If you roll 3, that female has three offspring. Now flip a coin three times to determine the sex of the three offspring: heads=female, tails=male. Under the next generation column record the three offspring, using the mother's corresponding mtDNA label and the sex of the offspring (i.e., if the mother has mtDNA label B, then one possible outcome based on the coin flips for the three offspring she produces is B-female, B-female, and B-male).
- If you roll 4, that individual has four offspring. Now flip a coin four times to determine the sex of the four offspring: heads=female, tails=male. Under the next generation column record the four offspring using the mother's corresponding mtDNA label and the sex of the offspring.
- If you roll 5, that individual has five offspring. Now flip a coin five times to determine the sex of the five offspring: heads=female, tails=male. Under the next generation column record the five offspring using the mother's corresponding mtDNA label and the sex of the offspring.

Keeping track of dispersal and reproduction for 30 woggles in three demes is quite a task, but we have provided tables for tallying your results as your simulation progresses. Follow the sample run of a single generation of the simulation illustrated in Tables 13.1, 13.2, and 13.3 before proceeding.

Test Statistic and Graphing Results

We want to examine the distribution of the five alleles across the three demes over a period of ten generations. We expect that gene flow will influence this distribution but we need to quantify the effect.

A test statistic (RR) was designed to quantify the distribution of alleles across demes. Low RR values mean that the demes share most of the alleles present. For example, if all three demes have the same alleles present, then the *R*-value will be 0. High RR-values mean that the demes share very few of the alleles present. You need to evaluate this test

TABLE 13.1.
Sample Deme 1.

Gen. 1	Dispersal or not (dice rolls)	Dispersal direction (coin flip)	Reproduction (dice rolls)	Sex (coin flip)	Gen. 2	Etc.
A-f	1, 5 (no disp.)		2 (no pups)			
B-f	5, 1 (no disp.)		6 (no pups)			
C-f	3, 2 (no disp.)		6 (no pups)			
D-f	1, 1 (dispersal)	Heads (counterclockwise)	Listed in deme 3 gen. 1			
E-f	4, 6 (no disp.)		4 (4 pups) ⟶	Heads	E-f	
A-m	1, 3 (no disp.)		(don't roll)	Tails	E-m	
B-m	2, 5 (no disp.)		(don't roll)	Tails	E-m	
C-m	1, 3 (no disp.)		(don't roll)	Heads	E-f	
D-m	2, 2 (dispersal)	Tails (clockwise)	Listed in deme 2 gen. 1			
E-m	5, 6 (no disp.)		(don't roll)			
A-f	From deme 3		6 (no pups)			
$D_1 = 5$					1	

TABLE 13.2.
Sample Deme 2.

Gen. 1	Dispersal or not (dice rolls)	Dispersal direction (coin flip)	Reproduction (dice rolls)	Sex (coin flip)	Gen. 2	Etc.
A-f	3, 5 (no disp.)		4 (4 pups) ⟶	Tails	A-m	
B-f	3, 2 (no disp.)		2 (no pups)	Tails	A-m	
C-f	1, 2 (no disp.)		2 (no pups)	Tails	A-m	
D-f	2, 1 (no disp.)		1 (no pups)	Heads	A-f	
E-f	5, 4 (no disp.)		3 (3 pups) ⟶	Tails	E-m	
A-m	4, 4 (no disp.)		(don't roll)	Heads	E-f	
B-m	2, 1 (no disp.)		(don't roll)	Tails	E-m	
C-m	6, 3 (no disp.)		(don't roll)			
D-m	4, 2 (no disp.)		(don't roll)			
E-m	4, 3 (no disp.)		(don't roll)			
D-m	From Deme 1		(don't roll)			
$D_2 = 5$					2	

TABLE 13.3.
Sample Deme 3.

Gen. 1	Dispersal or not (dice rolls)	Dispersal direction (coin flip)	Reproduction (dice rolls)	Sex (dice rolls)	Gen. 2	Etc.
A-f	2, 2 (Dispersal)	Heads (clockwise)	Listed in Deme 1 Gen. 1			
B-f	3, 2 (no disp.)		1 (no pups)			
C-f	3, 2 (no disp.)		3 (3 pups) ———→	Heads	C-f	
D-f	4, 5 (no disp.)		4 (4 pups)	Heads	C-f	
E-f	5, 4 (no disp.)		6 (no pups)	Tails	C-m	
A-m	2, 4 (no disp.)		(don't roll)	Heads	D-f	
B-m	3, 2 (no disp.)		(don't roll)	Tails	D-m	
C-m	6, 4 (no disp.)		(don't roll)	Tails	D-m	
D-m	5, 2 (no disp.)		(don't roll)	Heads	D-f	
E-m	5, 4 (no disp.)		(don't roll)			
A-f	From Deme 1		3 (3 pups) ———→	Heads	D-f	
				Heads	D-f	
				Heads	D-f	
$D_3 = 5$					3	

statistic *every generation* for the metapopulations at each level of dispersal. The equation for RR is

$$RR = (T - D_1)^2 + (T - D_2)^2 + (T - D_3)^2, \tag{13.1}$$

where T = number of different alleles represented in the entire metapopulation (range between 0 and 5); D_1= number of different alleles represented in deme 1 (range between 0 and 5); D_2= number of different alleles represented in deme 2 (range between 0 and 5); and D_3= number of different alleles represented in deme 3 (range between 0 and 5).

Note: You will tally these D values for each deme in each generation as you run the simulation; they are not the dispersal values (d) you use in the dispersal simulation rules! The RR for the example presented in tables 13.1–13.3 is presented in table 13.4.

Each student is the class is responsible for completing one run of the simulation for one level of dispersal. As you run the simulation you should gain a feel for the impact of dispersal and reproduction in these small populations. However, broader conclusions can only be drawn from examination of the pooled results of all of the simulations run by you and your classmates.

TABLE 13.4.
Sample calculation of RR statistic.

	Gen. 1	Gen. 2	Etc.
$D_1 =$	5	2	
$D_2 =$	5	3	
$D_3 =$	5	3	
$T =$	5	4	
RR=	$(5-5)^2 + (5-5)^2 + (5-5)^2 = 0$	$(4-1)^2 + (4-2)^2 + (4-2)^2 = 17$	

Questions to Work on Individually Outside of Class

Examine the example with three demes. Also examine how the number of alleles in each deme, and in all demes, are used to calculate RR for each generation. Now complete the three-deme simulation for the particular rate of dispersal assigned by your teacher. If he/she did not assign specific d-values, use the following guide: if your first name begins with the letters A–F use $d=0$; if G–L should use $d=1/18$; if M–R should use $d=2/18$; and if S–Z use $3/18$. Note that if a deme goes extinct, it can become recolonized. Record the results of your individual simulation in tables 13.5, 13.6, and 13.7. Also calculate RR for each generation and record these results in Table 13.8.

1. Plot the relationship between T (the number of different alleles present in the total metapopulation) on the y-axis, and generation (from 1 to 10) on the x-axis.

2. Plot the relationship between RR (the allelic distribution test statistic) and generation (from 1 to 10).

3. How did T change with generation for your metapopulation?

4. How did RR change with generation for your metapopulation? Did RR eventually reach a stable value? If not, would you eventually expect a stable value?

5. What potential biases may be created by this method of determining allelic distributions? How is RR affected by extinction in one or more of the demes?

You must turn in two graphs, your responses to questions, and your data tables with all data recorded, to receive full credit. You may work with friends but all data generated must be yours alone. Two people *may not* turn in the same data!

Instructions

Bring blank graph paper and your results to class; you will need these to complete the next part of this exercise.

Small-Group/In-Class Exercise: Save the Woggles

Secretary of the Interior A. R. Templeton has asked for a final report based on your simulations. He is interested in understanding how dispersal can counter the effects of genetic drift and extinction at the level of the deme. Specifically, he is interested in understanding

TABLE 13.5.
Worksheet for Deme 1 Simulation.

Gen. 1	Dis.	Gen. 2	Dis.	Gen. 3	Dis.	Gen. 4	Dis.	Gen. 5	Dis.	Gen. 6	Dis.	Gen. 7	Dis.	Gen. 8	Dis.	Gen. 9	Dis.	Gen. 10
A-f																		
B-f																		
C-f																		
D-f																		
E-f																		
A-m																		
B-m																		
C-m																		
D-m																		
E-m																		
$D_1=$		$D_1=$		$D_1=$		$D_1=$		$D_1=$		$D_1=$		$D_1=$		$D_1=$		$D_1=$		$D_1=$

TABLE 13.6.
Worksheet for Deme 2 Simulation.

Gen. 1	Dis.	Gen. 2	Dis.	Gen. 3	Dis.	Gen. 4	Dis.	Gen. 5	Dis.	Gen. 6	Dis.	Gen. 7	Dis.	Gen. 8	Dis.	Gen. 9	Dis.	Gen. 10
A-f																		
B-f																		
C-f																		
D-f																		
E-f																		
A-m																		
B-m																		
C-m																		
D-m																		
E-m																		
$D_2=$		$D_2=$		$D_2=$		$D_2=$		$D_2=$		$D_2=$		$D_2=$		$D_2=$		$D_2=$		$D_2=$

TABLE 13.7.
Worksheet for Deme 3 Simulation.

Gen. 1	Dis.	Gen. 2	Dis.	Gen. 3	Dis.	Gen. 4	Dis.	Gen. 5	Dis.	Gen. 6	Dis.	Gen. 7	Dis.	Gen. 8	Dis.	Gen. 9	Dis.	Gen. 10
A-f																		
B-f																		
C-f																		
D-f																		
E-f																		
A-m																		
B-m																		
C-m																		
D-m																		
E-m																		
$D_3=$		$D_3=$		$D_3=$		$D_3=$		$D_3=$		$D_3=$		$D_3=$		$D_3=$		$D_3=$		$D_3=$

TABLE 13.8.
Calculation of RR statistic.

	Gen. 1	Gen. 2	Gen. 3	Gen. 4	Gen. 5	Gen. 6	Gen. 7	Gen. 8	Gen. 9	Gen. 10
$D_1=$										
$D_2=$										
$D_3=$										
$T=$										
RR=										

how various degrees of habitat fragmentation, represented by various rates of dispersal, can influence the distribution of genetic variation between demes. Organize your report around the sections below.

Pooling Simulation Results

The secretary is aware of the fact that a single simulation run is meaningless, and as a result, he has asked everyone in your office to run the simulations. Each member of your group completed an independent simulation with the same dispersal rate (d). Pool your results and calculate the mean RR at generation 10. Enter that value in table 13.9 with the corresponding d-value. The other groups completed the simulations with each of the other three dispersal rates (d) listed at the top of the table. Please share your average RR value with the other three groups and ask them for the average RR values they got for the level of disper-

TABLE 13.9.
RR values for metapopulations with different migration rates.

d	0	1/18	2/18	3/18
Average RR at 10 generations				

sal they simulated. Record those values in your table as well. After you are done everyone should have the same pooled results to work with.

Dispersal and Allelic Distribution (Questions 6–9)

6. What is the biological significance of the relationship between d and RR?

7. Use the pooled results in table 13.9 and graph the relationship between dispersal (d) on the x-axis, and allelic distribution after 10 generations (RR) on the y-axis.

8. Did populations or demes go extinct in this simulation? What is the relationship between d and population size at the end of the simulation?

9. If each deme were adapted to very specific local microhabitat conditions, and these conditions differed across the demes, what would be the impact of dispersal on local adaptation?

Conservation Effort and Application (Question 10)

Given your familiarity with woggles and experience with the population dynamics of woggles, Secretary Templeton has asked your team to make specific proposals for preserving this species. As you know, woggles are currently restricted to a small subset of their historic range. There are three major metapopulations of woggles in California: Berkeley, Bakersfield, and Mojave. All three populations were sampled last year, and based on mark-recapture data all are estimated to be approximately the same size (100 animals). The three metapopulations also occupy approximately the same size area (800 sq. km). The demes east of Berkeley are relatively close to one another, and there are high rates of dispersal between demes. The metapopulation near Bakersfield runs along a dry valley, and dispersal has only been observed between adjacent demes. Most of the animals in the Mojave metapopulation are found in one valley but there are a few very small satellite demes (figure 13.1).

You have been asked by Secretary Templeton to assess the long-term viability of each of these three populations. Human development has increased in the areas surrounding each of these populations, and as a consequence, habitat fragmentation has increased further for each metapopulation. For each metapopulation you should consider the position and size of demes with respect to one another when making your assessment.

Secretary Templeton has awarded your group a new supercomputer to run more complex simulations. He has asked you to design simulations that specifically model the three existing metapopulations in California.

10. Describe the rules of your new simulation that can take into account the specific differences among these particular metapopulations. Describe all of the parameters in your simulation and why they are relevant. Explain the simulation rules and what they simulate. What other management practices would affect the metapopulations? Will these affect demographic extinction rates, gene extinction rates or both? How would you modify your model to incorporate these practices and predict the outcome?

Reference

Wintle, B. A., S. A. Bekessy, L. A. Venier, J. L. Pearce, and R. A. Chesholm. 2005. Utility of dynamic landscape metapopulation models for sustainable forest management. *Conservation Biology* 19(6): 1930–1943.

Section IV

Quantitative Ecological Tools

14

Understanding Descriptive Statistics

Beth Sparks-Jackson and Emily Silverman

Introduction and Background

Descriptive statistics synthesize and summarize data sets. They highlight important characteristics of a collection of measurements, like the most common or typical values, as well as average differences among or between individuals. Descriptive statistics include both single-valued summary measurements and graphical displays of data. The calculation of descriptive statistics for a large data set is the first step in any quantitative analysis; plotting the data is also critically important for understanding statistical results, comparing populations or groups of individuals, and checking the appropriateness of the statistical procedures being considered.

Basic Concepts and Terminology

Fundamental to descriptive and inferential statistics is an understanding of the terms *population* and *sample*.

The population is the entire collection of basic units and associated measurements about which information is desired. Examples of populations include:

Fecundity (offspring/year) of female lions in the Serengeti
Dollars per person contributed to conservation organizations from 1990 to 2000
Diameter at breast height and height class of trees infected by the emerald ash borer (*Agrilus planipennis*)

We can see that the statistical definition of a population is broader than the term's usage in other disciplines. For example, contrast the statistical definition of population with this common biological definition: a group of organisms of the same species populating a given area. Statistical populations may include one, two, or more measurements on each unit, known as univariate, bivariate, and multivariate data, respectively. The population may be finite and concrete (e.g., the heights of white oak trees in Central Park, New York City), or infinite and conceptual (e.g., the slope of each spot on every hill in Central Park). We use the upper case letter N to indicate the size of a finite population, but even finite and concrete populations are constantly changing.

Occasionally we may be able to measure every individual or object in a population. For example, the Nuclear Regulatory Agency is charged with a yearly accounting of spent fuel at all nuclear power plants in the United States. In most situations, however, measuring every unit in a population is either impossible or unrealistic and we need to measure a subset of the population.

A sample is a subset of a population. The number of measurements in the sample, or sample size, is typically denoted by the lower-case letter n.

Sampling is necessary when we cannot measure every individual in a population or when the measurement technique destroys the item being measured (for example, measuring mercury concentrations in fish brain tissue). Sampling can also save time and money. Ultimately we wish to generalize about a population from a sample; therefore, the sample must faithfully represent the population. Whether a sample is representative of the population depends largely on the sampling technique. Our examples here illustrate a few sampling methods. Sampling is explored further in Chapter 16.

How do we describe aspects of statistical populations? We use *parameters*, *estimates*, and *frequency distributions*.

Parameters are quantitative descriptions of the aspect of concern of a population. For example, the maximum possible, or largest, value is a population parameter.

Estimates, quantitative descriptions of a sample, are guesses at parameter values. For example, the maximum value in a sample is a guess at the population maximum.

Frequency distributions list all possible values of the measurement and the frequency or likelihood with which values are observed. Frequency distributions may be presented as tables or graphs. Sometimes frequency distributions can be described using mathematical functions. For example, the normal distribution represents distributions described by a specific bell-shaped curve.

Collectively, descriptive statistics help us describe and summarize key characteristics of a population. If our sample is representative of the population, our estimates and the distribution of the sample should be similar to the parameters and distribution of the population. For example, if the typical length of lizards in a given population is 10 cm, then lizards from a representative sample should also typically be around 10 cm long. If the population frequency distribution of lizard lengths is symmetric and hump shaped, the frequency distribution of a representative sample of lizard lengths should also be symmetric and hump shaped.

Types of Data and Common Representations

What do we do with sample data? How we present and analyze data largely depends on the type of data. This section explains how to identify data types (sometimes referred to as "scales of measurement") and describes graphical and nongraphical methods to present data.

Qualitative data

Qualitative, or categorical, measurements include two data types, nominal and ordinal. Nominal data can be distinguished only by type, state, or status. Examples of nominal data include:

Types of oak trees: red, white, pin, burr
Bird behavior: feeding, preening, flying, singing, sleeping
Gender: male, female

Ordinal data are also categorical, but the categories can be ordered. The distance between categories, however, is not necessarily equal and the measurements do not represent numbers even if labeled numerically. Examples of ordinal data include:

Measures of relative abundance: dominant, abundant, occasional, rare
Measures of quality: excellent, good, fair, poor
Birth order: first, second, third, etc.

When assigning data types, it may help to consider which mathematical functions apply. Equality is the only mathematical function that applies to nominal data; individuals are either the same type, state, or status or they are not. Because ordinal measurements are ordered, the applicable mathematical functions for these data include not only "equal to" (=) but also "less than" (<), and "greater than" (>). Nominal and ordinal data are often presented as counts; this can trick one into believing the data are numeric. Even though counts are numeric, the measurement itself is still categorical. For example, the gender distribution of 100 survey respondents may be presented as "47 males and 53 females" but, before being summarized as counts, the list of observations read "male, female, female, male, male, female, female," etc.

Qualitative data are frequently summarized as counts or percentages, and are usually presented in written, tabular, or graphical form. The simplest summary of qualitative data is the mode.

The mode is the value or category that occurs most frequently in the sample, i.e., the value or category with the largest count. If two values are equally abundant, the data are bimodal and both modes should be reported. The frequency distribution of qualitative data can be reported as counts or percentages in text, tables, and bar charts (box 14.1). Bar charts are graphical frequency distributions for categorical data and should be differentiated from histograms (graphical frequency distributions for numeric data) by leaving space between the bars. The data in bar charts should be ordered by frequency or some meaningful or natural order, such as genetic similarity or ordinal rankings (see figures 14.1 and 14.2).

Methods for summarizing, presenting, and analyzing categorical data are limited. We still, however, must choose carefully how to present and analyze these data. For example, although pie charts can be used to present the distribution of categorical data, problems arise when too many categories are included. A few categories might be presented best in the text or a short table. For these and other reasons, pie charts are not commonly used in the scientific literature. Including information such as the sample size (n) makes percentages for categorical measures more informative. Imagine a report stating that 33% of female lions pursued prey in a 24-hour observation period. If $n = 30$, we know that observations on many units went into the percentage estimate and we will be more confident in the correctness of the estimation of the population parameter than if $n = 3$.

Quantitative Data

Quantitative, or numerical, data employ a measurement scale in which the distance between observations is meaningful. There are two types of quantitative data: interval and ratio. Ratio data are much more common. Although interval data are rare, it is important to identify them because they cannot be analyzed using common estimators and statistical tests. Examples of interval data include orientation (compass degrees 0 – 360), time of day and dates, heights of tides, longitude, and temperature in °C or °F (not Kelvin, because ab-

Box 14.1 *Example presentations of categorical data*

These tabular data on aquatic species introductions are provided by the Food and Agriculture Organization of the United Nations.

Text: Of the 3141 recorded species invasions, only 20% have known ecological impacts. Of these species, 48.9% have known adverse effects, 29.7% are beneficial, and the direction of the effects of 21.3% are still unclear.

Table: Frequency distribution of types of invasive species, $n = 2574$.

Type of invasive species	% of records
Fishes	81.9
Molluscs	9.4
Crustaceans	6.0
Algae and Plants	1.1
Other invertebrates	0.9
Other vertebrates	0.7

Figures 14.1 and 14.2 are bar charts representing invasive species by taxon and causes of introduction.

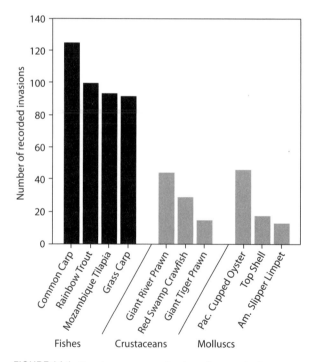

FIGURE 14.1. Bar chart representing invasive species by taxon.

(continued on following page)

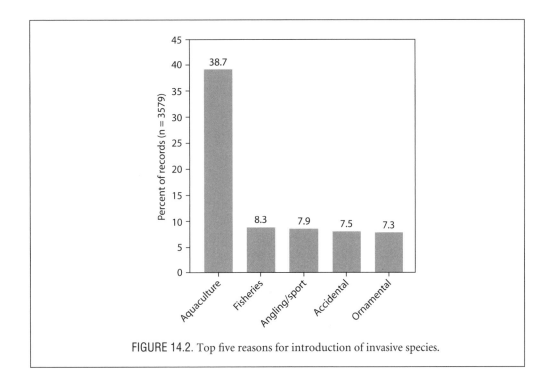

FIGURE 14.2. Top five reasons for introduction of invasive species.

solute zero is meaningful in Kelvin). Interval data have no true zero; further, many interval scales are circular (e.g., time, longitude, compass direction). The "zero" is often arbitrarily assigned and does not indicate absence. Addition and subtraction of interval data makes sense, but multiplication and division do not. An increase of 10°F represents the same amount of added heat from 34°F to 44°F and from 65°F to 75°F. Applicable mathematical functions are $=$, $<$, $>$, $+$, and $-$.

Ratio data are quantitative measurements with a true zero. Examples of ratio data include lengths of lizards, number of offspring, and distance seeds dispersed from their parent plant. Another way to distinguish this type of data is to consider whether ratios make sense. For example, a 20-year-old is twice as old as a ten-year-old. In contrast, it makes no sense to say that 62°F is "twice as hot" as 31°F. All mathematical functions ($=$, $<$, $>$, $+$, $-$, $\times$, and $\div$) apply to ratio data.

Ratio data come in two forms, continuous and discrete. Different presentation techniques and statistical analyses may be appropriate for these two types of ratio data. Continuous measures are defined by the property that between any two values there is always another possible value. Consider the measurement of human weight. Even though we commonly refer to weight in pound or kilogram increments, with a perfect scale we could theoretically record weights to an infinite number of decimal places. Discrete measurements, on the other hand, can only take certain values and may or may not include zero or negative values. Examples of discrete measures include number of offspring, number of limbs, and species richness. Discrete measurements may be manipulated to give conceptual (but not achievable) numbers (e.g., 2.5 children per household). Keep in mind, however, that continuous measures are reported as discrete because the precision of our instruments limits our ability to measure small differences.

Many techniques are available to describe and present quantitative data, including numeric measures of central tendency and spread. Measures of central tendency describe the value around which observations tend to cluster, i.e., the general location of a distribution on the *X*-axis. Measures of central tendency provide numerical answers to the questions "where are the data centered?" and "what are typical values?" Measures of spread or variability describe how similar the data are to one another and inform us of the range of likely values.

The two most common measures of central tendency are the arithmetic mean and the median. The arithmetic mean is calculated by summing all observations and dividing by the number of observations. Remember that we wish to guess at population parameters using estimates from samples. The population mean is the parameter μ (pronounced "mew") and is estimated by the sample mean $\bar{x}$ (pronounced "x-bar"). The sample mean is an unbiased estimator, meaning that, on average, $\bar{x}$ guesses μ correctly. The median is the center of the data when the observations are ordered from smallest to largest. If n is odd, the median is the central observation, and if n is even, the median is the average of the two central observations. An example calculation of the mean and median are included in box 14.2. The mode is rarely used to describe the central tendency of numeric data. The mode is particularly ineffective at describing continuous data, because every observation is a mode! We do not expect duplicate values or ties with continuous data, although some ties will occur due to the imprecision of our instruments.

Box 14.2 *Example Calculations of Measures of Central Tendency and Spread*

Dr. Phoebe Quayle is interested in comparing the growth rates of Western Gull chicks (*Larus occidentalis*) in colonies on Auklet and Birdman Islands. Phoebe recorded the growth rates (g/day) for 15 chicks for each of two colonies.

Let y = the growth rate in g/day of chicks on Auklet Island and let $n = 15$.

Observed growth rates are listed from smallest to largest:

$i =$ 1   2   3   4   5   6   7   8   9   10   11   12   13   14   15.
$y_i =$ 14.8 17.3 17.9 18.0 19.0 19.3 21.2 21.6 22.4 22.7 23.1 23.2 25.1 25.3 27.6.

Median m: Since $n = 15$, the median is the value for the central observation = $y_8 = 21.6$ g/day.

Sample mean, $\bar{y} = \sum_{i=1}^{n} y_i \Big/ n = (14.8 + 17.3 + 17.9 + 18.0 + 19.0 + 19.3 + 21.2 + 21.6 + 22.4$
$+ 22.7 + 23.1 + 23.2 + 25.1 + 25.3 + 27.6)/15 = 318.5/15 = 21.2$ g/day.

First quartile Q_1: The $0.25(n+1)$th = 4th largest observation = 18.0 g/day.

Third quartile Q_3: The $0.75(n+1)$th = 12th largest = 23.2 g/day.

Interquartile range (IQR): $Q_3 - Q_1 = 23.2 - 18.0 = 5.2$ g/day.

Sample variance, $s^2 = \sum_{i=1}^{n} (y_i - \bar{y})^2 \Big/ n - 1 = [(14.8 - 21.2)^2 + (17.3 - 21.2)^2 + (17.9 - 21.2)^2$
$+ \ldots + (25.1 - 21.2)^2 + (25.3 - 21.2)^2 + (27.6 - 21.2)^2]/(15 - 1)$
$= (40.96 + 15.21 + 10.89 + \ldots + 15.21 + 16.81 + 40.96)/14$
$= 170.19/14 = 12.16$ g^2/day^2

Sample standard deviation, $s = \sqrt{s^2} = \sqrt{12.16} = 3.49$ g/day.
Sample coefficient of variation, CV $= 100 \times s/\bar{y} = 100 \times 3.49/21.2 = 16.4\%$.

Which should we report, the mean or the median? This depends largely on the distribution and characteristics of the data. If the distribution is symmetric, the mean is equal to the median. A distribution is symmetric if a vertical line can divide the frequency distribution of the data into mirrored halves (figure 14.3a). If the distribution is skewed, or lacking symmetry, the mean and median will differ. A right-skewed distribution has an extended right "tail" (mean > median, figure 14.3b) and a left-skewed distribution has an extended left tail (mean < median, figure 14.3c). Because the mean makes use of every observation, it is sensitive to small and large observations. The mean shifts in the direction of the skew, while the median is less influenced by unusually large or small values of x (often called "outliers"). The median, therefore, is a more appropriate descriptor of the "typical" value when the data are extremely skewed.

The mean is more stable than the median; that is, the mean changes less as samples change. However, sometimes the median can be calculated when the mean cannot. Consider an experiment studying the survival times of 100 organisms. To calculate a sample mean, we must wait for all 100 organisms to die. To calculate the median, we only have to wait for 51 to die, since the median is calculated from the central observation(s). The median can also be used if the data are not measured with the same precision. Consider the following 11 abundance observations: 32, 43, 57, 64, 81, 88, 91, 98, >100, >100, >100. Since an abundance less than 100 is counted precisely, but anything over 100 is recorded imprecisely, only the median can be calculated.

Individuals in a population are never all identical (if they were, the measurement would be uninteresting and uninformative). Measures of central tendency do not describe how

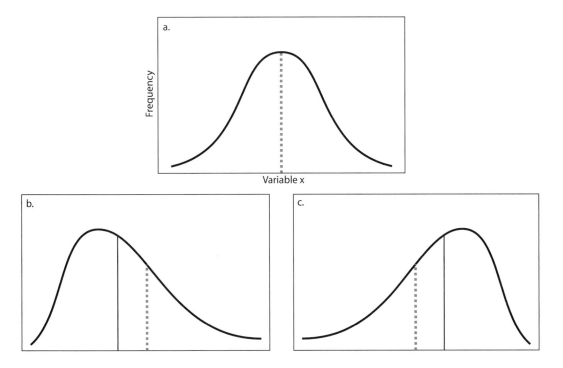

FIGURE 14.3. Skewed and symmetric distributions illustrating the relative location of the mean (dashed line) and median (solid line): (a) symmetric (no skew); (b) Right skew; (c) Left skew.

individuals differ. Measures of spread or variability allow us to gauge just how typical our "central value" is. One measurement of spread is calculated from the quartiles. Quartiles do just what their name implies: they split the sample into quarters. Q_2 is the median and splits the data into equal halves. The lower quartile, Q_1, separates the lower 25% of the sample from the rest and the upper quartile, Q_3, separates the upper 25% of the sample from the rest. In other words, Q_1 is the median of the lower half of the observations and Q_3 is the median of the upper half of the observations. Mathematically, Q_1 is the value of the $0.25(n+1)$th observation, and Q_3 is the value of the $0.75(n+1)$th observation, when observations are ordered from smallest to largest. The interquartile range (IQR), Q_3–Q_1, is a measure of the spread of the middle 50% of the data and is only minimally influenced by outliers. Quartiles and the IQR do not make use of every observation; they provide only a partial answer as to just how typical our central value is.

The variance and standard deviation are the most common measures of spread. The sample variance, s^2, estimates the population variance, σ^2 ("sigma squared"), and the sample standard deviation estimates the population standard deviation, σ. The sample variance, s^2, is the sum of squared deviations from the mean divided by $(n-1)$:

$$s^2 = \frac{\sum_{i=1}^{n} (x_i - \bar{x})^2}{n-1}. \tag{14.1}$$

Sample standard deviation, s, is the positive square root of the sample variance (Box 14.2).

Division by n seems natural, so why is $(n-1)$ in the denominator? Although we wish to estimate the spread of the observations around the population mean, μ, we don't know μ and must estimate it using the sample mean. The variance estimate assumes that the sample mean is the true population mean. This constraint (that the deviations come from a population with the observed sample mean) costs us one piece of information from our sample (called a "degree of freedom") leaving us with only $(n-1)$ pieces to estimate the variance.

Because the variance essentially measures the average squared deviation from the mean, the variance is small if all observations are similar to the mean, and large if observations tend to differ substantially from the mean. Extreme values, such as outliers, contribute substantially to s^2. The standard deviation measures the average deviation from the mean in the original units of measurement, unlike s^2 (square units). Thus, s is the most widely reported measure of spread.

The coefficient of variation, CV, expresses the standard deviation as a percentage of the mean (Box 14.2):

$$\text{CV} = 100 \cdot \frac{s}{\bar{x}}. \tag{14.2}$$

A unitless measure of variability is useful when we need to compare the variability of different groups of organisms, especially if σ^2 increases as μ increases, which is common. For example, consider the variability of mouse weights compared to the variability in elephant weights. We expect the elephants' standard deviation to be larger than the mice's standard deviation, but calculating the CVs allows us to compare the variability relative to the typical size of each organism. Box 14.2 provides example calculations of quartiles, IQR, s^2, s, and CV.

Numeric data can also be presented graphically, most commonly in histograms or box-plots. Histograms are graphical representations of the frequency distribution of a continuous variable. Rectangular bars are drawn so their bases lie on a linear scale representing equally spaced intervals, referred to as "bins." A bar's height is proportional to the frequencies of the values within each of the intervals (figure 14.4; numbers on the X-axis indicate the midpoint of a bin's range, e.g., the first bin includes observations between 7.75 and 8.25). Histograms allow us to examine our distribution for modality, skew, and gaps in the distribution (figure 14.5).

Although continuous measurements do not typically have single modes, "humps" in histograms suggest a typical range of values and result in distributions being described as unimodal, bimodal, or multimodal depending on the number of "humps." Histograms suffer from the fact that the apparent shape of the distribution may change depending on the number of bars or bins used (figure 14.6). Outliers are also sometimes difficult to identify in histograms.

A boxplot is another graphical technique for numeric data. Boxplots display information about the central location and variability of data. Most importantly, they give an indication of the symmetry of the data and the presence of possible outliers. The boxplot's "box" illustrates the range of the middle 50% of the observations. The median is the central line in the box, and the first and third quartiles are the outer boundaries of the box. A vertical line, often referred to as a whisker, connects the upper quartile to the most extreme observation within $1.5 \times IQR$ (interquantile range). A similar vertical line is drawn down from the lower quartile. Observations located beyond the vertical lines are indicated with a symbol, usually a star, dot, or line (figure 14.7).

Many hypothesis tests (see chapter 15) require data to be normally, or approximately normally, distributed. The normal distribution, described in more detail in the next section, is a particular symmetric, bell-shaped curve. Since critical departures from normality often occur because of skewed distributions or problems with the tails of a distribution,

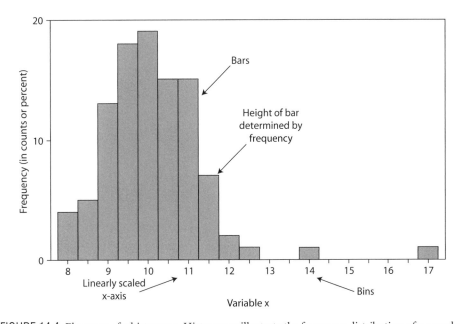

FIGURE 14.4. Elements of a histogram. Histograms illustrate the frequency distribution of a sample.

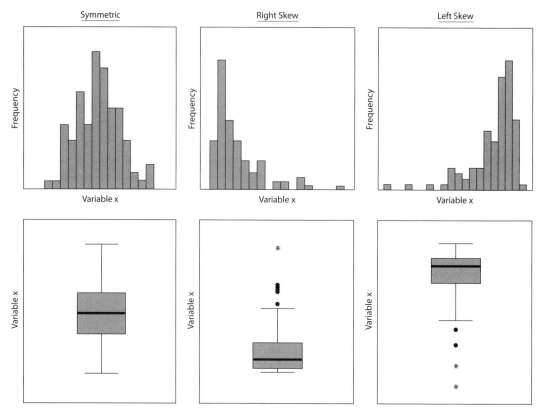

FIGURE 14.5. Examples of symmetric and skewed histograms and boxplots. For comparison purposes, the same data are presented in the corresponding histogram and boxplot.

a boxplot can be a useful diagnostic tool. However, the modality of a distribution can be obscured in boxplots, so data should be graphed in multiple ways during initial data exploration (figure 14.8). The location of the median in the boxplot indicates whether the data are symmetric or skewed. For example, if the median is closer to the lower quartile than the upper quartile, the data are somewhat right-skewed. The whiskers can give similar information (see examples of symmetric and skewed boxplots in figure 14.5). If the data are normally distributed, we expect to see an outlier only 0.07% of the time, or about one in 1000 observations. The presence of more outliers in a symmetric distribution suggests that some observations are too extreme for the data to be normal. The opposite problem, too few extreme values, is impossible to see in a boxplot. Boxplots also allow us to compare easily several distributions simultaneously, a difficult task with histograms.

Common Characterizations of Frequency Distributions

We can further describe data by characterizing their frequency distribution. Real data exhibit an infinite number of distributions, but if we look at many distributions, several gen-

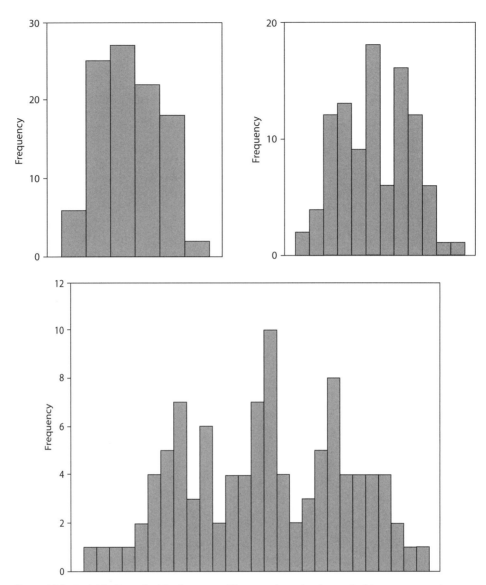

Figure 14.6. Variable *X* graphed in three ways illustrates how the shape of a histogram may change with the number of bins; e.g., 6, 12, or 30.

eral shapes and patterns emerge. These general shapes can be described by specific mathematical functions. Using one of these functions to describe the frequency distribution of a data set serves to simplify our description of the shape and characteristics of the data.

The normal distribution is the most commonly used distribution in statistics. Many continuous variables have distributions that resemble the normal, such as distributions of heights and weights. The normal distribution is actually a family of distributions whose particular locations and shapes vary with their mean μ and standard deviation σ (figure 14.9a). All normal distributions are unimodal and symmetric, but every bell-shaped dis-

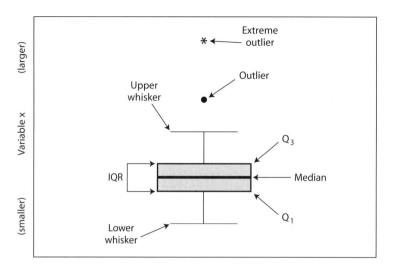

FIGURE 14.7. Elements of a boxplot. Boxplots also illustrate the distribution of a sample. They high-light skew and very large and small values.

tribution is not necessarily normal. The normal distribution has a specific arrangement of observations, or weight, in the tails of the distribution. If data are normally distributed, 68.3% of the observations fall within one standard deviation on either side of the mean, 95.5% within two standard deviations, and 99.7% within three standard deviations. The lognormal distribution is a particular right-skewed distribution that can be made normal by taking the logarithm of the original observations. The lognormal distribution is commonly used in studies of diversity to describe the distribution of the abundance of species in a given location.

Many other functions can be used to describe frequency distributions. Uniform distributions, sometimes referred to as "rectangular" distributions, are symmetric but differ from the normal in that each value has the same frequency of occurrence (figure 14.9b). The uniform distribution can be used to describe the pattern of discrete variables, e.g., the day of the week on which one was born, and continuous variables, e.g., the wait time for a bus if you arrive at a random time and the bus comes at set intervals. Poisson distributions model the distributions of discrete random variables. Commonly, the Poisson distribution is used when the variable of interest is a count of the number of rare events that occur in a certain time interval or spatial area; for example, the number of bird songs in 15 minutes or the number of alarm calls by prairie dogs within a 100 m^2 area. The shape of the Poisson distribution varies with the degree of rarity, λ ("lambda"), starting quite right-skewed for very rare events and becoming more symmetric as the counts increase (figure 14.9c). Exponential distributions model continuous variables with constant failure rates, such as fish survival times. The exponential is a severely right-skewed distribution where the smallest observations are most frequent and larger observations are relatively rare (figure 14.9d).

Summary and Presentation of Bivariate Data

Earlier in this chapter we mentioned that a population may consist of a set of units with multiple measures on each unit. When each sampling unit is associated with two mea-

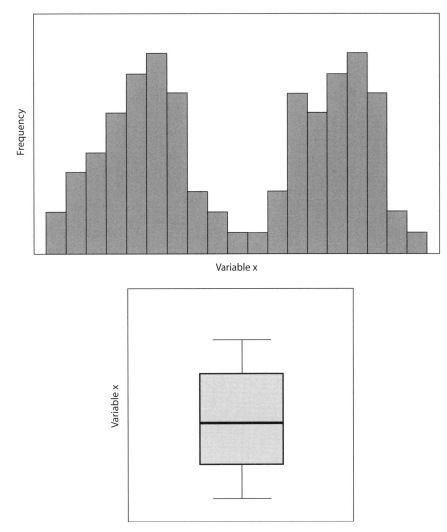

FIGURE 14.8. Variable *X* presented in a histogram and a boxplot. This illustrates a problem with boxplots: the data are clearly bimodal but the boxplot does not show this.

surements, the data are said to be bivariate. Although all of the descriptors and graphical techniques mentioned thus far can be applied separately to each set of measurements, new graphical and numeric techniques are also available. Bivariate data are plotted in a two-dimensional display called a scatterplot. In a scatterplot, points representing each measurement unit are displayed in a space defined by axes representing the two variables of interest. Scatterplots illustrate the relationship between the two variables *x* and *y* (figures 14.10 and 14.11).

The relationship between *x* and *y* can be summarized using either correlation or regression. Although these two techniques are sometimes used interchangeably, a clear distinction exists and the choice of correlation versus regression is not arbitrary. With correlation,

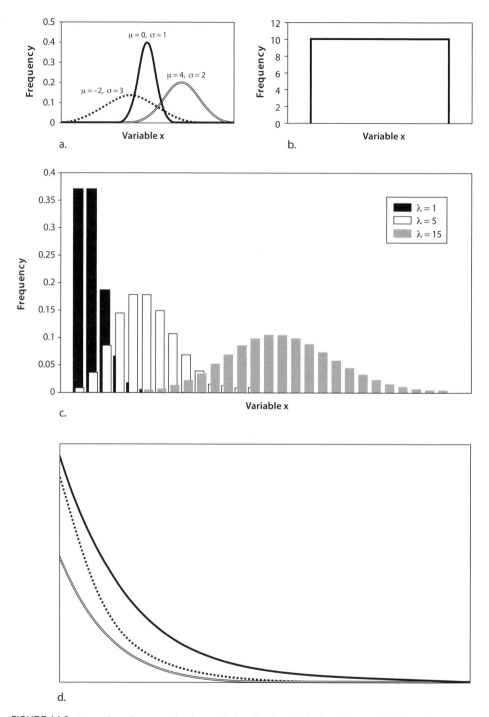

FIGURE 14.9. Examples of commonly observed distributions include: (a) normal, (b) continuous uniform, (c) Poisson, and (d) exponential. These mathematical distributions describe the frequency distributions of many ecological measurements.

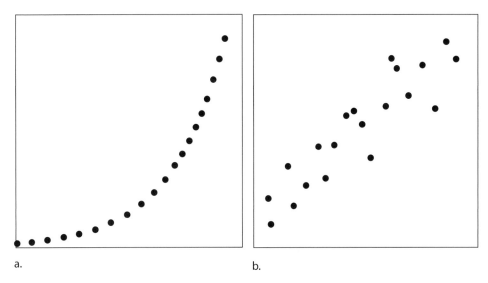

a.   b.

FIGURE 14.10. Even though there is a perfect relationship between x and y in (a) ($y = 0.88e^{0.30x}$), $r = 0.91$ in both (a) and (b). $n = 20$ for each plot.

we study how x and y covary. These two variables are interchangeable, and no dependence of y on x is implied. In regression, x and y are not interchangeable; x is the predictor, or explanatory, variable and y is the response variable. The goal of regression is to describe the variability of y in terms of the associated x value. Using regression, we estimate the linear relationship between variables by expressing one variable (y) as a linear function of the other (x).

In correlation, we quantify the relationship between x and y using the correlation coefficient. The population correlation coefficient ρ ("rho"), is estimated by r, the sample correlation coefficient, and describes the tightness and direction of the linear association between x and y. Correlation does not accurately describe nonlinear relationships (figure 14.10). The correlation is not the slope of the relationship between x and y. Values for r range from $+1$, a perfect positive linear relationship between x and y, to -1, a perfect negative linear relationship. An r of zero indicates no linear relationship at all. Several example scatterplots and their associated correlations are illustrated in figure 14.11. In regression, we quantify the relationship between x and y by estimating the slope (β) and y-intercept (α) of the line ($y = \alpha + \beta x$) that best fits the data. The regression line describes how y changes linearly as x changes. How is the "best fit" line determined? The most common approach is least squares regression. Least squares regression chooses a line that minimizes the sum of the squared vertical distances between the line and each point (figure 14.12). The regression line will always passes Through $\bar{x}$ and $\bar{y}$.

The coefficient of determination, R^2, is used in regression analyses to describe the relationship between x and y. R^2 is the proportion (or percentage) of the variability in y that can be explained by the linear relationship between x and y. If reported as a proportion, R^2 ranges from 0 to 1; if reported as a percentage, the range is 0 to 100. If $R^2 = 0$, then none of the variability in y is explained by x and if $R^2 = 1$ (or 100), then all of the variability in y can be explained by x. Another way to think about this is to consider the "utility" of the

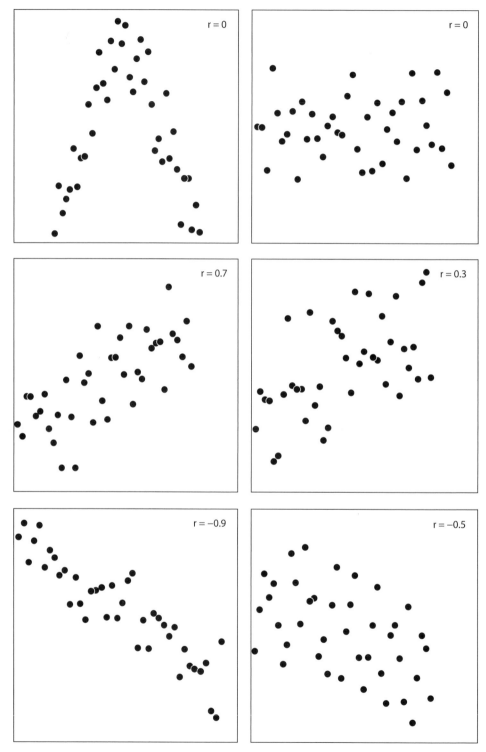

FIGURE 14.11. Various scatter plots and associated *r*. *n* = 40 in each plot.

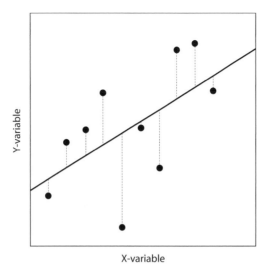

FIGURE 14.12. Illustration of least squares regression. The line is chosen to minimize the sum of the squared distances, indicated by vertical lines.

equation of the best fit line. An R^2 of 0 tells us that knowing x gives us no information about the expected values of y, while an R^2 of 1 (or 100) indicates that, by using the relationship between x and y, y can be predicted exactly for each observed x. Although it is mathematically true that $R^2 = (r)^2$, the two terms should not be used interchangeably, just as correlation and regression are not interchangeable. R^2 is the appropriate descriptive statistic for regression while r is the appropriate descriptor for correlation. Figure 14.13 provides several examples of regression lines and R^2's.

Homework for this exercise takes approximately one to two hours.

Objectives of this Exercise

In this exercise you will practice identifying types of data, calculating and interpreting sample estimates, and creating and interpreting textual, tabular, and graphical displays of data.

Case Study

The Western Gull (*Larus occidentalis*) is a large gull that nests along most of the Pacific coast in the United States and southern Canada. Western Gulls are common and populations are now stable, although pesticides and human persecution were once a threat. Many populations recovered after the automation of U.S. Coast Guard light stations and the closing of Alcatraz Federal Penitentiary, since personnel at these facilities caused extensive seabird disturbance. Western Gull reproductive ecology and feeding are well studied. Although these gulls feed primarily on fish, their diet can also includes refuse, carrion, aquatic invertebrates, and the eggs and chicks of other birds.

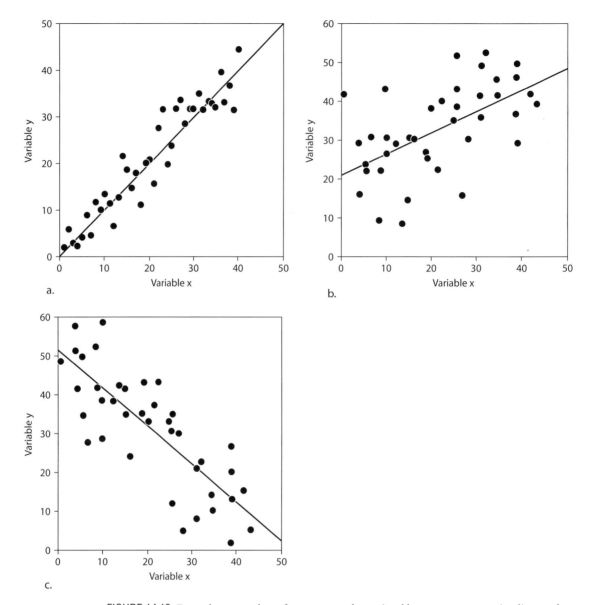

FIGURE 14.13. Example scatter plots of y versus x and associated least squares regression lines and R^2 values. $n = 40$ in each plot. (a) $y = 0.52 + 0.98x$ and $R^2 = 0.90$ or 90%; (b) $y = 21 + 0.55x$ and $R^2 = 0.34$ or 34%; (c) $y = 51 - 0.97x$ and $R^2 = 0.66$ or 66%.

Questions

Your professor, Dr. Phoebe Quayle, studies the reproductive success of Western Gulls. To earn a bit of extra spending money you have decided to join Dr. Quayle's research team this summer. For the past few months you have been reading journal articles to familiarize yourself with Western Gull ecology such as Sydeman et al. (1991). You have also met with Phoebe and she has provided some unpublished data for you to consider. Please answer the following questions.

Questions to Work on Individually Outside of Class

1. One of the variables that Sydeman et al. (1991) were interested in measuring was the availability of juvenile shortbelly rockfish (*Sebastes jordani*), the preferred food for Western Gull chicks on southeast Farallon Island. One way to measure rockfish abundance is to pump the stomachs of king salmon (*Oncorhynchus tshawytscha*), a predator of juvenile rockfish. The amount of rockfish found in salmon stomachs was assumed to be a good proxy for the amount of rockfish available to foraging Western Gulls. Propose a situation where a sample of salmon stomachs might be representative of the rockfish available to Western Gulls. Propose a situation where the salmon proxy would likely *not* be representative of the rockfish available to Western Gulls.

2. During your literature review you found the following measures in studies of Western Gulls. Identify each type of data as nominal, ordinal, interval, or ratio. If the measure is numeric, identify whether it is discrete or continuous. Justify your answers.

   (a) Reproductive history (recorded as: none, built nest but no young fledged, 1 chick fledged, at least 2 chicks fledged)
   (b) Typical direction of parental foraging flights (in degrees, with north set at 0)
   (c) Number of eggs in nest
   (d) Chick fledge date
   (e) Prey species fed to chick
   (f) Growth rate of chicks (g/day)
   (g) Identification number on banded adults

3. Sydeman et al. (1991) measured the age and timing of breeding of over 3000 Western Gull pairs. They pooled breeding age into young (3–5 years), middle-aged (6–12 years), and old (13–21 years) age groups and they grouped clutch initiation time into early, peak, and late (see table 14.1).

   (a) Before the authors pooled the data, they were originally recorded as age in years and date of clutch initiation. What kind of data were these original measures? What kind are they now that the data are pooled?
   (b) Create a bar chart of age at nest initiation with age on the *X*-axis and different bar shadings representing the three timings of nest initiation. (Unless you are very familiar with a computer graphing program, just draw this, and other plots, by hand.) From your chart, does it look as if the age at reproduction is related to timing of nest initiation? Why or why not?

4. Last summer, Phoebe recorded the most abundant and second most abundant types of food in the stomachs of 20 adult Western Gulls (table 14.2). Since she hasn't had time to summarize and present the data, she has asked you to do this. Present these data in the three different forms for categorical data discussed in this chapter (text, table, bar chart). Choose which of these you feel is the best way to present these data. Justify your choice.

5. In one of Phoebe's previous studies you found the histogram (figure 14.14); you want to use the data in another analysis. Unfortunately, Phoebe recently moved offices and all of her old data files are packed away.

TABLE 14.1.
Breeding time for young, middle-aged, and old Western Gulls.

	Young	**Middle-aged**	**Old**
Early	68	592	212
Peak	272	1085	249
Late	227	296	69
$n =$	**567**	**1973**	**530**

Note: Counts per class estimated from figure 1 in Sydeman et al. (1991).

(a) Look at the histogram carefully. Is it necessary for Phoebe to unpack her box of disks for you to have the original data? Explain.

(b) Calculate the mean and median number of eggs per Western Gull nest.

(c) Based on your calculations and the shape of the histogram, describe the skew of the distribution.

6. Earlier in this chapter we presented a sample of growth rates from the colony of Western Gulls on Auklet Island. Box 14.2 gives the growth rates from Birdman Island. For this colony, calculate the sample mean, median, 1st and 3rd quartiles, IQR, s^2, s, and CV.

TABLE 14.2.
"Raw" stomach content data for 20 Western Gulls.

Gull ID	Most abundant prey	Second most abundant prey	Gull ID	Most abundant prey	Second most abundant prey
1	Fish	Eggs/nestlings	11	Fish	Human refuse
2	Human refuse	Fish	12	Aquatic invertebrates	Fish
3	Aquatic invertebrates	Fish	13	Fish	Human refuse
4	Fish	Human refuse	14	Aquatic invertebrates	Eggs/nestlings
5	Human refuse	Fish	15	Human refuse	Fish
6	Carrion	Aquatic invertebrates	16	Fish	Aquatic invertebrates
7	Fish	Human refuse	17	Fish	Human refuse
8	Human refuse	Aquatic invertebrates	18	Eggs/nestlings	Carrion
9	Fish	Aquatic invertebrates	19	Fish	Human refuse
10	Fish	Carrion	20	Fish	Carrion

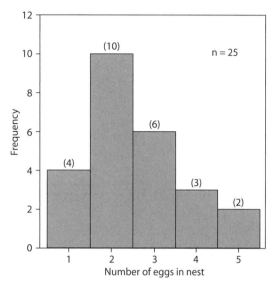

FIGURE 14.14. Dr. Quayle's old data on clutch size.

Growth rate (g/day) (n = 15, listed in order of collection): 18.6 20.2 20.4 23.6 22.0 24.8 17.6 15.3 19.1 16.3 13.2 16.1 24.3 14.8 19.9.

7. You would like to compare the average growth rate of chicks from Auklet and Birdman Islands, but you notice that the procedure you want to use requires the data to be normally distributed. Using the data from the two groups and some of the calculations from question 6 and box 14.2,

(a) create two histograms, one for each sample (since $n = 15$, you will need to limit yourself to just a few bins);
(b) create a figure with two boxplots representing the distributions of the data from these two locations;
(c) from these figures, justify whether you think the data are normally distributed.

8. During your literature review, you found figure 14.15 describing the distribution of breeding life spans for over 100 pairs of Western Gulls during a 12-year study on Alcatraz Island, California (reproduced from Annet and Pierotti, 1999).

(a) How would you describe this distribution? Does it resemble any of the mathematical distributions discussed in this chapter?
(b) What does this tell you about the age distribution of breeding pairs of Western Gulls?

9. In the same article you found another figure (figure 14.16) that examines the relationship between parental diets and diets of male offspring (reproduced from Annett and Pierotti, 1999).

(a) In the article, the authors state: "diets of young males during the first year of breeding were correlated with the diets of their parents ($r^2 = 0.404$)." Given the differences between regression and correlation, what is wrong with this statement?

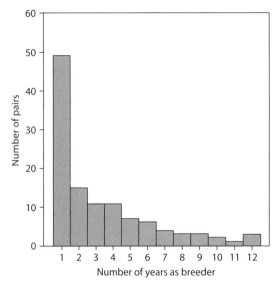

FIGURE 14.15. Breeding tenure of Western Gulls on Alcatraz Island.

(b) Although the article does not present the equation for the least squares regression line, you estimate it to be: Percent of fish in male offspring's diet = −10.7 + 0.89 × percent of fish in parent's diet. Draw this line on the figure. Using a different style of line (e.g., dashed), draw in the 1:1 line.

(c) What does the location of observations relative to the 1:1 line tell you about male offspring diets compared to that of their parents?

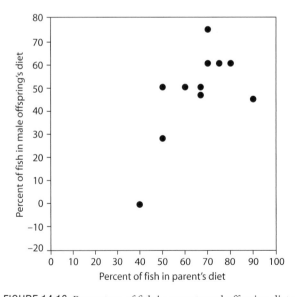

FIGURE 14.16. Percentage of fish in parents and offspring diets.

(d) Circle the two points that contribute the most to the low R^2 value.

(e) Describe how the fit and slope of the regression equation would change with the removal of either of these observations and with the removal of both observations.

Small-group / In-Class Exercise

Note that chapters 14 and 15 are connected. You will use your answers to these problems in the following chapter.

Understanding Nest Predation by Western Gulls

Now that Western Gull populations are doing so well, some ecologists have become concerned about the potentially harmful effects of nest predation by Western Gulls. Dr. Phoebe Quayle has invited you to join her research team on a new project studying the predation of Western Gulls on the eggs and nestlings of other birds. In particular, you are interested in comparing predation intensity on Auklet and Birdman Islands. You are also interested in the factors that contribute to the frequency of predacious feeding events. Phoebe has asked you to brainstorm ideas in preparation for an upcoming meeting in which the research design will be decided. Consider the following questions.

Research Question 1: Does the Intensity of Predation by Western Gulls Differ Between the Colonies on Auklet and Birdman Islands?

10. Obviously you need a measure of predation intensity, but there are many different ways to obtain this.

(a) There are two "basic units" on which predation intensity can be measured: (1) observing how much the adult Western Gulls eat or (2) observing the nests that are subject to predation. For each of these two approaches, identify a nominal, ordinal, and ratio measure of nest predation intensity.

(b) Which, if any, of these measures are observational and which involve physical handling of the birds?

(c) Are there reasons for choosing one measure over another? Consider statistical and logistical issues.

11. Choosing each of the measures you listed in question 10, explain how you might present your data in a scientific journal article.

12. Choose a ratio measure of predation from question 10. Sketch a hypothetical figure assuming that the measure is normally distributed and on average the same on the two islands, but predation is more variable on Auklet Island.

13. Sketch a figure assuming your ratio measure is right-skewed and equally variable for both locations, but that the average is larger for Birdman Island.

Research Question 2: What Factors Might be Associated with Frequency of Nest Predation by Western Gulls?

14. Assume you choose to measure predation intensity (y) by capturing adult Western Gulls and recording grams in the predator's stomach due to nest predation. Identify a ratio and nonratio measure (x) that may be associated with predation intensity (y).

15. Considering the ratio measure you chose in question 14, is this measure positively or negatively related to predation intensity? Is correlation or regression more appropriate for describing this relationship? Justify your answer.

16. You think that predation on nestlings and eggs may be more frequent if the nests are clustered. In a pilot study you measure 15 individuals on Auklet Island and estimate the relationship between the distance in meters to a nest with available nestlings/eggs (x) and grams in the predator's stomach (y) to be $y = 45 - 0.38x$ with $R^2 = 0.8$. Sketch this relationship and the best fit line.

17. You also measure the same variables for 15 individuals on Birdman Island and estimate the relationship between x and y to be $y = 32 - 0.2x$ with $R^2 = 0.4$. Draw this line on your figure from question 16. Using the y-intercept, the slope, and the coefficient of determination, describe the differences in the relationship between the two islands.

References

Annett, C. A., and R. P. Pierotti. 1999. Long-term reproductive output in Western Gulls: consequences of alternate tactics in diet choice. *Ecology* 80:288–297.

Sydeman, W. J., J. F. Penniman, T. M. Penniman, P. Pyle, and D. G. Ainley. 1991. Breeding performance in the Western Gull: Effects of parental age, timing of breeding and year in relation to food availability. *Journal of Animal Ecology* 60:135–149.

15 Understanding Statistical Inference

Emily Silverman and Beth Sparks-Jackson

Introduction and Background

Throughout this book, we examine scientific questions and their translation into hypotheses that generate testable predictions. But exactly how do we decide whether a hypothesis is supported by the data or whether we should reject it? Statistical inference is the process of drawing conclusions about population characteristics from sample data. This process includes assessing uncertainty due to the fact that a sample will never represent a population exactly. Inference may be used to answer questions about a single population, compare two or more populations, or explain the relationship between two different measurements from a single population. Statistical inference includes (i) the testing of hypotheses about populations, and (ii) the estimation of population parameters using sample data. (See Example, Part 1 at the end of the case study and data section.)

Homework for this exercise takes approximately one hour.

Objectives of This Exercise

This exercise will familiarize you with the basic principles of traditional statistical inference and improve your understanding of published scientific results. The discussion questions are based on Phoebe Quayle's data for Western Gulls, which you explored in the chapter 14 exercises.

Basic Concepts and Terminology

Hypotheses

Statistical tests about populations begin with two statements: the null and alternative hypotheses (see box 15.1).

When conducting a hypothesis test, we ask, "Is there convincing evidence that the null hypothesis is wrong?" If we claim that something is different from the *status quo*, such as "The earth's climate is warming," then the hypothesis test says "Show me!"

There are two outcomes to any hypothesis test. If the data provide strong evidence that the null hypothesis is wrong, we reject the null hypothesis in favor of the alternative. Or, if

---

Box 15.1 The null and alternative hypotheses are two mutually exclusive statements about some characteristic of the population under study.

The **null hypothesis**, usually denoted H_0, is a specific statement about the population; it is typically the hypothesis of the *status quo* or no difference.

The **alternative** (or "research") **hypothesis**, usually denoted H_a or H_1, is nonspecific, stating that the null hypothesis is not true.

See Example, Part 2.

---

the data do not provide this strong evidence, we fail to reject the null hypothesis. Thus, *we can never prove a hypothesis*. We can only disprove one (the null). This is analogous to the fact that we can never prove something is safe; we can only disprove safety by demonstrating a specific harm.

Test Procedure

In order to make a decision about our null hypothesis, we will calculate the distance between H_0 and the sample data. This calculation is done using a *test statistic*.

The test statistic measures the difference between the sample data and the population as defined by the null hypothesis. This difference takes into account the fact that the sample estimates will not equal the population values, even when H_0 is true.

1. Large values of the test statistic are evidence that the H_0 is false. "Large" may be large and positive, large and negative, or large in absolute value, depending on the type of hypothesis test and the specific null hypothesis.
2. The distribution of the test statistic is known, if H_0 is true. This allows us to calculate the probability of observing the value of the test statistic that we did, or larger, if H_0 is true. This is the *p-value*.

    The *p*-value is the probability of a test statistic as large as, or larger than, we observed, given that H_0 is true. It measures the likelihood of the observed value of the test statistic, or a value more extreme, if H_0 is true.
3. If *p* is small, we have evidence against the H_0 and in favor of the alternative, H_a. See Example, Part 3.

Types of Error

All hypothesis tests are based on the same basic principles and are set up to control and minimize the probability of drawing an incorrect conclusion. There are two possibilities: the null hypothesis is correct or it is false. There are also two outcomes to a test: we reject the null hypothesis or we do not. Thus, there are two ways to get things right and two ways to make a mistake. This situation is summarized in table 15.1 with the shaded boxes indicating incorrect decisions.

A Type I error occurs when H_0 is true, but is rejected. The rate, or probability, of a Type I error is denoted by the Greek letter α.

A Type II error occurs when H_0 is false, but is not rejected. The rate, or probability, of a Type II error is denoted by β.

TABLE 15.1.

Types of error.

Reality ⟹ Decision ⇓	H_0 is true	H_0 is false, H_a is true
H_0 is rejected	α	$1-\beta$
H_0 is not rejected	$1-\alpha$	β

See Example, Part 4.

Reality is not negotiable; we are either in column 1 or 2 of the table. Because we do not know which column we are in, we must concern ourselves with both types of error, even though only one is truly possible. The test decision is mutually exclusive: we either reject the null hypothesis, or we don't. Thus, each column's values sum to one, i.e., some decision is guaranteed.

In hypothesis testing, we preset the probability of a Type I error, α, to a small value (often 0.05 or 0.01). This is called the significance level of the test.

Decision Rule

To determine whether we should reject our null hypothesis, we compare the p-value to α. We reject H_0 if $p < \alpha$ and do not reject H_0 if $p > \alpha$. This is our decision rule. By comparing the p-value to α, we ensure that we will commit a Type I error with rate α, when H_0 is true. See Example, Part 5.

Power of the Test

We would like to make both types of error, α and β as small as possible. But, if all else is equal, decreasing one error rate will increase the other: by making it hard to reject H_0 when H_0 is true, we make it harder to reject when H_0 is false.

Because we know the distribution of our test statistic if the null hypothesis is true, *we can control α, the probability of falsely rejecting H_0.* We cannot, however, specify β, the probability of not rejecting H_0 when it is false. This error rate depends on *how* H_0 is false, i.e., on a specific alternative. But, H_a says only that the null hypothesis is incorrect. Since the specific alternative is not stated, it is not possible to determine β exactly.

We try to minimize β by a number of indirect methods (discussed below). Minimizing β is equivalent to maximizing $(1 - \beta)$ or the probability of correctly rejecting H_0; $(1 - \beta)$ is called the power of the test. Power is a critical component of every hypothesis test. When the null hypothesis is not rejected, it is possible that we have failed to reject an incorrect H_0. If power is high (so β is low), we can be relatively confident we did not make this error. But, if power is low, our results are ambiguous: is there truly no evidence against H_0 or was the failure to reject H_0 simply due to low power?

How do we ensure high power? The power of a hypothesis test depends on four quantities:

1. The significance level α. The larger the value we choose for α, the higher the power. By increasing the chance of a Type I error, we decrease the chance of a Type II error. Decreasing β increases power $(1 - \beta)$.

2. The sample size n. The larger the sample size, the higher the power. This relationship should be intuitive: the more information we have from the population(s), the more likely it will be that we draw the correct conclusion.

3. The difference between H_0 and reality (i.e., H_a). Again, this should be intuitive: the further the null hypothesis is from the truth, the easier it should be to reject.

4. The population standard deviation, σ. The higher the population variability, σ, the lower the power. The more variable the population, the harder it should be to draw the correct conclusion.

See Example, Part 6.

Table 15.2 summarizes the hypothesis testing procedure used by Phoebe Quayle in the example.

Common Hypothesis Tests

The appropriate hypothesis test depends on the question and the type of data. If we have a hypothesis about a population distribution or are comparing two population distributions, then we should employ a goodness-of-fit test. For example, we might want to know if the time between pollinator visits to a flower follows an exponential distribution or if the age distribution of two populations of lions is the same. If we have a hypothesis about a population mean and our data are likely to be normally distributed, then a t-test is recommended. If the data are not normally distributed, then we would need to conduct an alternative test. Tests that do not require the assumption of normality are called nonparametric or distribution-free tests.

Consider an extension of take-home question 8, chapter 14, as an example: you might wish to test whether the distribution of breeding life spans on Alcatraz Island differed between the 1990s and recent years (2000 to the present). In this case, you are testing a hypothesis that compares the distributions of two populations, and the data are quantitative; therefore you could use the Kolmogorov-Smirnov test. However, since the time of observation for recent years is less than the original study (which lasted 12 years), you would likely group all breeders with more than six years breeding (assuming the "present" is 2005), treat the data as categorical, and use a contingency table test for homogeneity.

Hypothesis tests concerning population parameters, like the mean, variance, and population correlation, can be one- or two -sided. This means we can test whether a parameter equals a specific value, or is larger or smaller than a specific value. Tests comparing distributions (and chi-square contingency tests for relationship) are typically non-directional, i.e., we simply test equality versus difference and do not consider the direction of the difference.

If the samples are large, the tests listed in table 15.2 that require the data to be normally distributed are relatively robust to deviations from this assumption. This means that, if the data are not normal but the samples are large, the tests still do well. The non-parametric tests listed are rank tests, because the original measurements are replaced by their ranks. Note also that the Mann-Whitney U test and Wilcoxon signed-rank test are sometimes referred to by other names. Finally, there are procedures available to test hypotheses about population variances. Variance tests that make the assumption of normality are not robust to deviations from normality, and must be applied with caution. These hypothesis tests, and many others, are described in most introductory statistic texts. Two useful statistical reference texts used by ecologists are *Biostatistical Analysis* by J. H. Zar and *Biometry* by R. R. Sokal and F. J. Rohlf.

TABLE 15.2.
Common statistical tests and their uses.

To test a hypothesis about …	Data are …	Test is …
Distributions		
The type of distribution of a single population *or* Comparing the distributions of two populations/groups	Categorical	χ^2 goodness-of-fit test Contingency table χ^2 test for homogeneity
The type of distribution of a single population *or* Comparing the distributions of two populations/groups	Quantitative	Kolmogorov-Smirnov test
Means		
The value of the mean of a single population	Quantitative and normal	One-sample *t*-test
	Quantitative and non-normal *or* ordinal	Wilcoxon signed-rank test
Comparing the mean of two populations/groups	Quantitative and normal, samples independent	Two-sample *t*-test
	Quantitative and non-normal, *or* ordinal, samples independent	Mann-Whitney *U* test
	Quantitative and normal, samples paired	One-sample *t*-test
	Quantitative and non-normal *or* ordinal, samples paired	Wilcoxon signed-rank test
Comparing the means of three or more populations/groups	Quantitative and normal, samples independent	One-way analysis of variance
	Quantitative and non-normal, samples independent	Kruskal-Wallis test
Bivariate relationships		
Association between two variables	Categorical data	Contingency table χ^2 test for independence
Association between two variables	Quantitative and normal (association assumed linear)	*t*-test for Pearson's correlation coefficient
	Quantitative and non-normal *or* ordinal	Test for Spearman's rank correlation
The values of regression line's *y*-intercept and slope	Quantitative and normal	One-sample *t*-test

Confidence Intervals

In some situations, we collect sample data to estimate population parameters rather than to test a specific hypothesis. Examples include:

1. Abundance estimation entails sampling to determine the total number of plants or animals, τ "tau"), in some area of interest.
2. To set fishing limits, fisheries biologists might wish to know the mean daily fish catch by anglers, μ ("mu"), in a certain river.

"Point" estimates, like the sample mean, sample median, and sample standard deviation, represent our best guess at the corresponding population values. But we can be relatively certain that the estimates we calculate from a sample are not equal to the correct values! Because the sample estimate is unlikely to hit the population value exactly, we may wish to use our sample data to estimate a range of likely values for the population parameter.

Confidence intervals are intervals that cover the population parameter with a specified probability. For example, [19.3g, 23.2g] is the 95% confidence interval on the mean growth rate of Western Gulls on Auklet Island, based on the sample data presented in box 14.2. This interval is calculated in such a way that the probability that it covers the true mean growth rate, μ, is 0.95. Repeated samples of size 15 from Auklet Island will each result in a slightly different confidence interval, but we would expect 95% of them to cover the colony's mean growth rate. Thus, a 95% confidence interval has an error rate of 5%. The endpoints of the interval are called the upper and lower confidence limits, and 0.95 is called the confidence coefficient (or level). Other common confidence levels are 0.99 and 0.90. A 99% confidence interval will always be wider than the 95% confidence interval for the same sample, because the 99% interval has a higher probability of covering the population mean; the 90% confidence interval will be narrower than the 95% interval.

The proper statistical procedure to calculate a confidence interval depends on the population parameter(s) of interest and the nature of the data (e.g., whether the observations are normal or not). In many cases, there is a one-to-one correspondence between a confidence interval and the results of a hypothesis test. For example, assume that we wish to test the null hypothesis that mean growth rate for Auklet Island is equal to a specific value (say μ_0), i.e., $H_0: \mu = \mu_0$, at $\alpha = 0.05$. Using the one-sample t-test, we will not reject this null hypothesis for any specific value, μ_0, that lies within the 95% confidence interval; we will reject H_0 for all values of μ_0 outside the interval. Thus, the confidence interval corresponds to a range of possible values of the population parameter, determined from the sample data and based on a specified error rate. It is also possible to create one-sided confidence intervals (e.g., lower or upper confidence bounds only); these correspond to one-sided hypothesis tests. See Example, Part 7.

The relationship between confidence intervals and hypothesis testing is important: p-values and the conclusion of "reject" or "fail to reject," by themselves, are not especially interesting or scientifically important. Results are best considered with respect to the range of likely values of the population parameter.

Example

Reconsider Dr. Phoebe Quayle's data on Western Gulls, *Larus occidentalis*, from chapter 14. Phoebe recorded growth rates in g/day for 15 chicks on each of two islands. The summary statistics are $\bar{y}_1 = 21.2$, $s_1^2 = 12.16$ and $\bar{y}_2 = 19.1$, $s_2^2 = 12.78$, where 1 = Auklet Island and 2 = Birdman Island.

1. Phoebe believes that the food availability differs between the islands and that this results in different chick growth rates for the two populations. She would like to compare the mean growth of chicks on Auklet Island, μ_1, to the mean growth on Birdman Island, μ_2.

2. Phoebe's null hypothesis is that there is no difference between the two islands: the mean growth rate on Auklet Island is the same as on Birdman Island. Mathematically, H_0: $\mu_1 = \mu_2$. Her alternative, or research, hypothesis is that mean growth rate differs between the two islands, or, H_a: $\mu_1 \neq \mu_2$.

If $\bar{y}_1$ is very different from $\bar{y}_2$, then Phoebe may have evidence that the null hypothesis is wrong. If so, she would reject H_0 in favor of H_a. The difference between the sample means is measured relative to the variability in the two samples (s_1^2 and s_2^2). A given difference between two sample means is more likely to indicate a difference between the population means if the variability of the populations (estimated by the sample variances) is small.

Phoebe's hypothesis is called "two-tailed" because her research hypothesis is supported if there is evidence that $\mu_1 > \mu_2$ or if there is evidence that $\mu_1 < \mu_2$. There are two possible one-tailed research hypotheses: (i) mean growth rate on Auklet Island is greater than on Birdman, or H_a: $\mu_1 > \mu_2$ and (ii) mean growth rate on Auklet Island is less than on Birdman, or H_a: $\mu_1 < \mu_2$. The corresponding null hypotheses are rejected only if there is evidence in the direction of the specific alternative (e.g., $\bar{y}_1 \gg \bar{y}_2$ for the first case).

3. If Phoebe is confident that her measurement, chick growth rate, is normally distributed, she would use the two-sample t-statistic to measure the distance between the data and the null hypothesis (see table 15.2). The formula for this statistic is

$$t = \frac{\bar{y}_1 - \bar{y}_2}{\sqrt{(s_1^2 + s_2^2)/15}}. \tag{15.1}$$

This statistic measures the difference between the sample means, relative to the average variability of the two samples. If H_0 is false, the t-statistic will be large and positive, or large and negative.

In this case, $t = 1.63$. The p-value for this problem is 0.11; this is the probability of observing t greater than 1.63 or less than −1.63 (a large positive or large negative is evidence against Phoebe's null hypothesis). Thus, the value of $t = 1.63$ is somewhat likely, if H_0 is true. Phoebe does not have strong evidence to suggest that mean growth rate differs between the two islands.

4. If Phoebe rejects her null hypothesis and concludes there is evidence that the mean growth rate differs in the two populations, when it actually does not, then she has committed a Type I error. Alternatively, if Phoebe fails to reject her null hypothesis and concludes there is no evidence of differences in growth between the two islands, when there in fact is a difference, then she has committed a Type II error.

5. Assuming that Phoebe sets her probability of a Type I error to be 5%, she will not reject H_0 because $p = 0.11 > 0.05 = \alpha$. Based on this result, Phoebe could not have committed a Type I error, although she may have committed a Type II error (if H_0 is false).

6. If $\alpha = 0.1$, instead of 0.05, Phoebe would still not reject H_0, but she would come closer to doing so. If the sample size from each population were 30, instead of 15, then $t = 2.41$ (larger) and $p = 0.02 < 0.05$. Phoebe would then reject H_0. If the difference between μ_1 and μ_2 were 2.9, instead of 2.1, then $t = 2.25$ and $p = 0.03 < 0.05$. Phoebe would again reject H_0. Finally, if the average sample variance were 10.0, instead of 12.47, $t = 2.74$ and $p = 0.01 < 0.05$, once again, Phoebe would reject. Decreasing α, increasing sample size, increasing the

difference between the populations (and hence samples), and decreasing the variability of the populations (and hence samples), increases the power of the test.

7. The 95% confidence interval for the mean difference in growth rate between the two islands is [−0.49, 4.79]. This interval includes zero, indicating that the null hypothesis that the two colonies have the same growth rate is not rejected at α = 0.05. This result matches the result in question five. The 99% confidence interval is [−1.41, 5.72]. The interval is wider because it must cover the true mean difference with probability 0.99, not 0.95.

Questions to Work on Individually Outside of Class

1. In take-home question 3 of chapter 14, you explored the relationship between gull age and the timing of nest initiation. You wish to determine whether the relationship between these two variables is statistically significant. Using table 15.3, answer the following questions:

(a) What is the type of hypothesis you wish to test?
(b) What type of data do you have?
(c) What is the appropriate test?

2. The p-value for the pair age/nest timing test in question 1 is less than 0.001.

(a) Is the null hypothesis rejected at α = 0.05?
(b) What is the apparent relationship between age and timing of nest initiation?
(c) Do you think the test had high or low power? Why?
(d) Would the power be higher or lower, if there had been 1000 pairs instead of around 3000? Explain.

3. Assume that Phoebe would like to test the research hypothesis that the mean number of eggs/nest is larger than two.

(a) Which section of table 15.2 includes the possible tests Phoebe might use?
(b) What type of data does Phoebe have? Based on the histogram from take-home question five, chapter 14, do Phoebe's data look normally distributed? What hypothesis test procedure in table 15.2 should she use? Defend your choice.
(c) Write down Phoebe's null and alternative hypotheses.
(d) The p-value for Phoebe's test is 0.02. If Phoebe set α = 0.05, would she reject the null hypothesis? Why?
(e) Given the test result in (d), what error could Phoebe have made? What error is not possible, given her test result?

TABLE 15.3.
Gull age and nest initiation.

	Young	Middle-aged	Old
Early	68	592	212
Peak	272	1085	249
Late	227	296	69
$n =$	567	1973	530

(f) Robin Jay, Phoebe's graduate student, believes that Phoebe should set $\alpha = 0.01$. If Phoebe followed Robin's advice, would she reject the null hypothesis? Given this test result, what error could Phoebe have made?

(g) Explain the difference between Phoebe's and Robin's results in (d) and (f).

4. Take-home question 9, chapter 14, presents the least squares regression line relating offspring diet to parent diet: Percent fish in male offspring's diet $= -10.7 + 0.89 \times$ percent fish in parent's diet. Consider the following results: "The slope of the line relating offspring diet to parent's diet was not significantly different from one ($t = -0.31$, $p = 0.76$); in addition, the y-intercept did not differ from zero ($t = -0.47$, $p = 0.65$)."

(a) Based on the hypothesis test results, is the percentage of fish in a male offspring's diet the same as the percentage in his parent's diet? Explain.

(b) The 95% confidence interval for the slope is [0.12, 1.67]. The 95% confidence interval for the y-intercept is [−62.22, 40.82]. Explain what these intervals tell you about the relationship between fish in the parent and offspring diets. Do these confidence intervals "match" the hypothesis test results? Explain.

(c) If the influential point (40,0) is removed from the analysis, the regression line is: Percent fish in male offspring's diet $= 25.73 + 0.39 \times$ percent fish in parent's diet. The value of the test statistic for the null hypothesis that the slope is equal to one, t, is now -1.88. In this case, is the p-value likely to be larger or smaller than for the full dataset (where $p = 0.76$)?

(d) The 95% confidence interval for slope, corresponding to the estimate in (c), is [−0.35, 1.14]. Based on this interval, would you reject the null hypothesis that the slope is equal to one at $\alpha = 0.05$? Would you reject the null hypothesis that the slope is equal to zero (no relationship between male offspring and parent diet of fish) at $\alpha = 0.05$? Explain your answer and discuss what the test can and can't tell you in this case. Do you think these tests have high or low power?

Small-Group/In-Class Exercise

Understanding Nest Predation by Western Gulls, Continued

Today you will revisit your work in chapter 14 in order to test hypotheses about Western Gull predation.

Research Question 1: Does the Intensity of Predation by Western Gulls Differ Between the Colonies on Auklet and Birdman Islands?

1. For each of your three measures of predation pressure (nominal, ordinal, and ratio) by the predator, answer the following questions.

(a) For each measure, what population characteristics (e.g., distribution, mean, etc.) can you compare between the two islands?

(b) State a null and an alternative hypothesis you would use to compare nest predation by Western Gulls on Auklet and Birdman Islands for each measure.

(c) Based on the type of hypothesis you stated in (b) and the type of measurement, state which hypothesis test procedure from table 15.2 you would use in each case and explain why.

2. Assume that you wished to test the research hypothesis that nest predation was worse on Birdman Island. For which measure(s) could you test this hypothesis? For which could you not? Explain your answer.

Research Question 2: What Factors Might be Associated with the Frequency of Nest Predation by Western Gulls?

3. Consider the two explanatory variables (x_1, x_2) you suggested might be associated with predation intensity (y) for in-class question 14, chapter 14.

(a) Which test from table 15.2 would you use to determine if the mean stomach contents of the predators (y) differs among the categories of your nominal (or ordinal) explanatory variables? Hint: Find the appropriate type of test in table 15.2 and consider the number of categories of the explanatory variable.

(b) Which test from table 15.2 would you use to determine if predation intensity (y) is related to your ratio explanatory variables? What assumption is necessary for this test to be meaningful?

4. In chapter 14, in-class question 16, you considered the relationship between y and the distance to the nearest nest (x). The least squares estimated linear relationship was $y = 45 - 0.38x$.

(a) What are the units associated with the estimated y-intercept? The slope?

(b) In take-home question 4 of this chapter, you compared the slope of the relationship to one. Is that a meaningful null hypothesis for the current variables? What is the null hypothesis of interest for here? Is it one or two-sided? Explain.

(c) Now consider all the information presented in in-class questions 16 and 17, chapter 14, and the null hypothesis you suggested in (b). For which island, Auklet or Birdman, do you expect to have a "larger" test statistic (i.e., relative to your null hypothesis and possibly considering "more negative" as larger)? Explain.

References

Sokal, R. R., and F. J. Rohlf. 1995. *Biometry: The Principles and Practice of Statistics in Biological Research*. New York: W. H. Freedman & Co.

Zar, J. H. 1999. *Biostatistical Analysis*. Upper Saddle River, N.J.: Prentice-Hall.

16 Sampling Wild Populations

Stanton Braude and James Robertson

Introduction and Background

Despite the impression we get from nature movies on television, we can rarely examine every individual in a wild study population. Instead, we sample some subset of individuals and extrapolate our results to the whole population (see chapters 14 and 15). Depending on the species, the structure of the habitat, and our research question, we can employ a variety of methods to select and analyze our sample. Samples can be systematic, simple random (see chapter 14) or stratified random (e.g., stratified by sex, so the two categories are male and female, and then sampled randomly within category). What is appropriate depends on the research question. Here we will explore random sampling techniques.

One method, quadrat sampling, is useful for many plants, bird nests, scat, or sedentary animals. In basic quadrat sampling, we lay a grid over our study site and select a subset of the squares to sample exhaustively. For example, in each quadrat, we might count the number of stems of every kind of grass. In radius quadrat sampling we select points within the study site and exhaustively sample areas of a given radius around those points. In transect sampling, we select a vector (or vectors) through the study site, and sample all individuals within a narrow area around that vector (e.g., 1 meter wide). A road may also serve as a transect; for example, road kills can be counted to estimate relative abundance of some species (but note that roads are not random!).

Mark-Recapture Sampling

For cryptic or shy animals, mark-recapture methods can be used to sample a population and estimate population size. Petersen's basic mark-recapture method (Petersen, 1894) depends on capturing a random subset of the total population each time traps are set.

If all animals trapped on one day are marked, the proportion of those marked animals that are retrapped on a later day should be proportional to the total number of animals in the population (with the important assumption that there are no deaths, births, immigration, or emigration in the meantime). Because total trapping success on subsequent days may differ (due to variation in weather, bait freshness, or a variety of other factors) the total population size in the trapping area is estimated by the equation:

$$\frac{\text{number of marked animals in the second sample}}{\text{total number of animals caught in the second sample}} = \frac{\text{number of marked animals in the first sample}}{\text{total population size}}. \tag{16.1}$$

Random Distribution

The Hopkins-Skellam (1954) method of assessing distribution of individuals in space is based on the following: if individuals are spaced randomly, the statistical distribution of distances from individuals to their nearest neighbors will be the same as a distribution of distances from random points to individuals.

The Hopkins-Skellam statistic $A = W_1 / W_2$. We first select a random series of points in the study area and measure the distance from each point to the nearest individual. W_1 is the sum of the squares of distances from these random points to the nearest individuals. Next we select a random subset of individuals in the population and measure the distance to each of their nearest neighbors. W_2 is the sum of the squares of distances from these individuals to their nearest neighbors. If $A = 1$, the distribution of individuals in space is random. If $A > 1$, the distribution is clumped. If $A < 1$, the distribution is regular or hyperdispersed. In this exercise we will not test statistically whether A is significantly greater or less than 1. However, you can discuss the distribution of individuals in space as "slightly more clumped than random" or "quite regularly dispersed," etc. when citing your calculated Hopkins-Skellam A value in your work for this chapter.

In this exercise you will use a variety of sampling methods to estimate population abundance, density, and spatial distribution. The sampling methods you employ here can be used to estimate other population parameters, including age, reproductive status, fertility (e.g., litter size, clutch size, or mast crop), parasite infestation, and population genetic parameters.

Homework for this exercise takes approximately one hour.

Objectives of This Exercise

In this exercise we will examine sampling data to compare the advantages and limitations of different methods. We will also critique real data, and design and execute our own sampling protocol.

Case Study and Data

Based on your extensive knowledge of field ecology, you have been offered a position with African Wildlife Travels (AWT). This organization designs and develops ecotourism travel packages throughout sub-Saharan Africa. A portion of AWT profits is directed toward animal and plant conservation through its subsidiary, AWT Conservation (AWTC).

As the new staff vertebrate ecologist (SVE), your first duty is to complete a survey of the abundance and distribution of several African vertebrate species in Kenya and Namibia (figure 16.1). Eventually you will design and execute sampling protocols for a wide range of vertebrate species. When designing a sampling method, you must, of course, consider the natural history of the organism and be aware of possible biases or weaknesses in the chosen method. A sampling method that is appropriate for one species, or a certain type of question, may not be appropriate for another species or type of question.

Before you begin your own work, you have to clean up the mess left by your predecessor, the previous staff vertebrate ecologist, who was asked to resign due to poor job performance. (Like so many of AWT's employees, he loved wildlife, loved being in the bush,

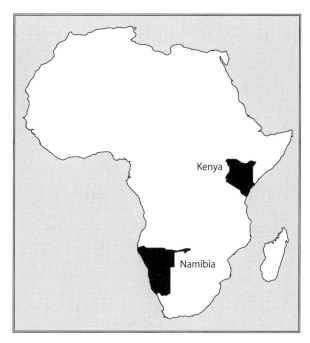

FIGURE 16.1. Areas sampled for this exercise. Kenya is located in the eastern horn of Africa and Namibia is located in southwest Africa.

and could amass reams of data, but he was worthless at quantitative analysis and critical interpretation.) He had spent over $68,000 on a six-month safari and collected data on three vertebrate species when he was called back to the head office and fired for failing to complete any of his reports.

You can assume that the actual data he reported are legitimate; however, you should reevaluate whether the appropriate sampling methods were used in each case. You should also consider any possible disadvantages of the sampling methods used. If a more appropriate method exists, you should discuss that method and its feasibility. Feel free to consult with experts or call up your old classmates from college.

Here are the raw data that your predecessor collected on three different African mammals before his forced resignation:

Mouse-Tailed Bat Rhinopoma kardwickei *(Figure 16.2)*

Identification: total length 13 cm, tail 6.5 cm, wing span 23 cm. A creamy brown bat easily distinguished by its very long, slender tail, which extends far beyond the poorly-developed interfemoral membrane. Appears at dusk; flight is slower and less erratic than most insectivorous bats (Williams, 1981).

Habitat: Roosts in colonies in caves; hunts insect prey flying over freshwater.

Geographic distribution: In East Africa, known from northern Turkana, Kenya, where it is recorded from Central Island and Ferguson's Gulf, Lake Rudolf, and at Lokomorinyang in the extreme north, and at Lake Baringo. Also occurs in northeastern Karamoja, Uganda (Williams, 1981).

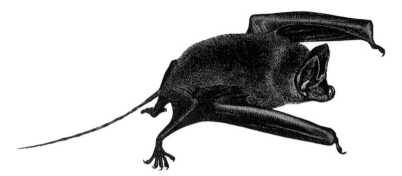

FIGURE 16.2. Mouse-tailed bat *Rhinopoma kardwickei*.

Sampling method and data: The assignment was to estimate population sizes in three different parks in northwest Kenya. Mist nets, finely woven nylon mesh fishing nets, were used to catch the bats. The nets were hung across several lakes within each park. At night the bats leave their roosts in caves to fly over the lakes and feed. They fly into the mist nets, where they remain tangled and trapped until the researcher returns to mark and release the bats. A transponder microchip with a unique coded number was implanted under the skin of each bat; this was scanned through the skin for future identification. Ten nets (each 3 m × 10 m in size) were used at each of the three parks.

Baringo:
January 20, total caught and marked=26
January 28, total caught=34 (20 of these were already marked)
Sibiloi:
February 14, total caught and marked =14
February 21, total caught=10 (3 of these were already marked)
Saiwa Swamp
March 1, total caught and marked=17
March 9, total caught=17 (9 of these were already marked)

Springbok Antidorcas marsupialis *(Figure 16.3)*

Identification: total length 1,200–1,400 mm; tail 150–300 mm; shoulder height 730–870 mm; weight 30–48 kg. Cinnamon fawn above, with a dark reddish brown horizontal band extending from the upper foreleg to the edge of the hip, separating the upper color from the white underside. Both sexes have black, ringed horns (Nowak, 1991).

Habitat: Lives on open, dry savannahs and grassland. The springbok is suspicious of roads and wagon paths, and sometimes clears these obstacles at a bound. Springbok are able to survive without water, though they will drink when water is available. They occur in large aggregations during the wet season, when males are territorial; they form smaller groups during the dry season. Drought occasionally forces the springbok to undertake large-scale migrations in search of new pastures.

Geographic distribution: Occurs in Angola, Namibia, Botswana, and South Africa.

Sampling method and data: Your predecessor was asked to estimate population sizes in two different parks in Namibia. Five randomly chosen transects were made in each park. The transects were 1 kilometer long and piles of fecal pellets located no more than 2 meters from the transect on either side were recorded. He reported the following data:

FIGURE 16.3. Springbok *Antidorcas marsupialis.*

Etosha National Park: (April 20–24)
    transect 1—33 fecal pellet piles
    transect 2—54 fecal pellet piles
    transect 3—45 fecal pellet piles
    transect 4—27 fecal pellet piles
    transect 5 —49 fecal pellet piles
Khaudom Game Reserve: (April 29–May 2)
    transect 1—22 fecal pellet piles
    transect 2—17 fecal pellet piles
    transect 3—29 fecal pellet piles
    transect 4—58 fecal pellet piles
    transect 5—31 fecal pellet piles

African Elephant Loxodonta Africana *(Figure 16.4)*

Identification: Largest living terrestrial animal. Size differs slightly according to sex. Adult females weigh 2160–3232 kg and range in shoulder height from 2.2 to 2.60 m. Adult bulls weigh 4700–6048 kg and range in shoulder height from 3.20 to 4.01 m. Hair sparsely covers large triangular-shaped ears. Noted for thick curved tusks (Laursen and Bekoff, 1978).

    Habitat: Elephants occur in all types of country, from mountain forest to semiarid bush and savanna. They can be found at altitudes from sea level to 3660 m and occasionally in snow at 4570 m.

    Geographic distribution: Can be found as far north as the Sudan. Westward, elephants occur in isolated areas around Lake Chad, in Mali, and Mauritania. They range south to South Africa where they are found in the Addo Elephant, Knysna, and Kruger National Parks. They no longer occur in most of southern South Africa (Namibia), Botswana, or

FIGURE 16.4. African elephant *Loxodonta africana.*

in Ethiopia or northern Somalia in northern Africa. They are abundant in Kenya, Zimbabwe, Tanzania, Zambia, Uganda, and Zaire (Democratic Republic of Congo) but they are mainly restricted to protected sanctuaries. These elephants are common in many national parks and game reserves and in other areas of East Africa. Very large tusked bulls occur on Marsabit Mountain in Northern Kenya.

Sampling method and data: Your predecessor was asked to estimate population sizes in two different refuges in Kenya. In each case the parks were divided into quadrats and 10% of the quadrats were randomly chosen for observation. Helicopters were used to fly over the chosen quadrats and all observed elephants were recorded for each quadrat. Many of the elephants are permanent residents within the particular refuge in which they were observed. The elephants, however, are not confined to the refuges and seasonal migrations are known.

Meru: 1,813 square kilometers total. Divided into 100 quadrats of approximately 18 sq. km per quadrat.

10 quadrats sampled on May 25 and 26

Quadrat 1–10: 7, 4, 12, 2, 0, 1, 0, 5, 6, 10 elephant observations.

Kora: 2,070 square kilometers total. Divided into 100 quadrats of approximately 21 sq. km per quadrat.

10 quadrats sampled on June 25 and 26
Quadrat 1–10: 4, 0, 2, 3, 5, 1, 2, 0, 6, 4 elephant observations.

Questions to Work on Individually Outside of Class

To complete your predecessor's first three projects, answer the following questions. Be sure to include the appropriate units (e.g., sq km) in all your calculations.

1. Can you estimate the abundance, density, and spatial distribution of the mouse tailed bat populations in Baringo, Sibiloi, and Saiwa Swamp? Calculate these estimates when possible. If you were unable to calculate these estimates, explain why. What assumptions did you make in order to calculate these estimates? (Show your work.)

2. Can you estimate the abundance, density, and spatial distribution of the Springbok populations in Etosha and Khaudom? Calculate these estimates when possible. If you were unable to calculate these estimates, explain why. What assumptions did you make in order to calculate these estimates? (Show your work.)

3. Can you estimate the abundance, density and spatial distribution of the elephant populations in Meru and Kora? Calculate these estimates when possible. If you were unable to calculate these estimates, explain why. What assumptions did you make in order to make these estimates? (Show your work.)

4. How would you have collected data differently? What additional data would you have collected, and why?

5. Based on your work with other chapters in this text, how might wildlife managers in Africa use these data to inform their decisions?

Small-Group/In-Class Exercise

Depending on weather conditions, collect data to evaluate one of these sampling problems:

Option A: Sampling Resources on Campus

Congratulations! Your team has an excellent reputation for your quantitative problem solving; we want you to apply your experience to planning and executing a sampling study of human resources on campus. Choose one of the following questions, and plan your sampling method before heading out of the classroom (table 16.1). Your tools may include a five-meter tape measure (provided by your instructor), a random number table, and possibly a map. Take approximately 20 minutes to plan your work and the next hour to gather your data and summarize your results.

Questions

1. Food. In our campus environment, food is mostly available in cafeterias and restaurants and from vending machines. Estimate the abundance and density of these resources on campus by sampling five quadrats. Are they randomly distributed?

TABLE 16.1.
Random number table.

9	9	8	7	8	8	6	0	3	5
1	9	1	4	3	5	1	5	0	5
8	4	3	6	5	2	8	5	5	8
7	0	0	6	4	7	2	0	2	9
4	3	7	4	8	7	5	7	8	2
3	7	3	0	7	0	1	8	5	1
2	0	8	5	1	9	2	4	6	4
1	2	5	1	2	3	9	8	4	0
0	9	1	2	4	1	0	5	8	7
3	0	2	8	6	0	4	4	9	9

2. Water. Drinking fountains in hallways are the primary source of this critical resource in our environment here on campus. Estimate the abundance of this resource on campus by sampling five buildings on campus and extrapolating.

3. Space. Although space for standing or walking may be abundant on campus, there is often competition for seating and for table study space. Estimate the abundance and density of seating or tables on campus by sampling five buildings and extrapolating. One of the candidates for student council has asserted that the school needs to provide more space for studying. Estimate the percent occupancy to test whether this resource is, in fact, limited.

Option B: Sampling Wildlife on Campus

Congratulations! Your team has an excellent reputation for your quantitative problem solving. Since the weather is so nice out today, you will be applying your experience to plan and execute an ecological sample on campus. Choose one of the following questions, and plan your sampling method before heading outside. Your tools will be a five-meter tape measure (provided by your instructor), a map of campus, and a random number table. Take approximately 20 minutes to plan your work and then the next hour to gather your data and write up your results.

Questions

1. Paving. Many students have complained that our once-beautiful campus is gradually being paved over. What proportion of the area on campus is covered with concrete, stone, or tarmac? (To be fair, exclude any area that is occupied by a building.)

2. Tree Cover. Two of the most beautiful old trees on campus blew down in a storm last year. After seeing them cut up and hauled away students have become concerned that campus is at risk of becoming barren because older trees are not being replaced by young trees. While you may not have the knowledge to identify trees by species, you can at least estimate the density of trees on campus and average DBH (diameter at breast height is a standard

used by foresters, defined as 4.5 feet above ground, measured from the uphill slope when the tree is growing on a slope).

3. Green Slopes. Graduation is approaching, and some of the university administrators are concerned about the muddy appearance of our campus. Although the recruiting brochures show students relaxing and reading on the green slopes, this inevitably leads to destruction of even the hardiest fescue grass. What proportion of the grassy areas on campus is actually covered with grass, and what proportion is bare ground that could become muddy on graduation day?

References

Hopkins, B., and J. Skellam. 1954. A new method for determining the type of distribution of plant individuals. *Annals of Botany* 18: 213–227.

Laursen, L., and M. Bekoff, 1978. *Loxodonta Africana. Mammalian Species* 92: 8.

Nowak, R. 1991. *Walker's Mammals of the World*, 5th ed. Baltimore: John Hopkins University Press.

Petersen, C. 1894. On the biology of our flat-fishes and on the decrease of our flat-fish fisheries. Report of the Danish Biological Station No. IV (1893–94).

Williams, J. G. 1981. *National Parks of East Africa*. Lexington, Mass.: Stephen Greene Press.

17 Quantifying Biodiversity

Cawas Behram Engineer and Stanton Braude

Introduction and Background

The term "biodiversity" has many different meanings. Definitions of biodiversity range from "the totality of genes, species, and ecosystems in a region" (WRI, 1992), to "the variety and variability among living organisms and the ecological complexes in which they occur" (UNEP, 1992), and "the diversity of life in all its forms, and at all levels of organization" (Hunter, 1996), to "the ensemble and the interactions of the genetic, the species and the ecological diversity, in a given place and at a given time" (Castri and Younes, 1996). This last definition encompasses the interactions within, between, and among the different levels of biodiversity (figure 17.1). This definition of biodiversity takes into account not only the species diversity, but also the diversity at the genetic and ecosystem levels.

Here we will use this definition: biodiversity is the sum total of all species (plant, animal, fungal, and microbial), along with their genetic variations, found in a particular habitat/ ecosystem; it includes the different interactions that go on at the genetic, species, and ecosystem levels. Biodiversity is loosely correlated with ecological stability or robustness, although for any particular case, causality may be unclear. That is, once there exists considerable diversity, chance extinctions of one taxon are unlikely to cause the entire system to collapse. However, we are unlikely to see great diversity emerge except in regions where conditions (e.g., climate) are moderate and stable and the area is not isolated—conditions that permit large numbers of immigration events.

Comparing biodiversities in different places is often important in setting conservation priorities. But how can we measure such a diverse set of parameters that will allow us to compare the biodiversity in different parts of the world?

Biodiversity can be characterized by parameters such as:

(a) Species richness: there are several definitions but they are based on the number of species in an area and equal importance is assigned to each species.

(b) Taxic diversity: measures the taxonomic distribution of species. Taxic diversity also highlights species that are evolutionarily isolated but are important in a given system.

(c) Functional diversity: evaluates the complex interactions among food webs, keystone species, and guilds in an area which provides a measure of richness in functional features, strategies, and spectra.

(d) Species diversity: measures the number of species in an area and takes into account sampling effects and species abundance.

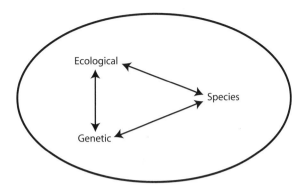

FIGURE 17.1. No matter what level of biodiversity you are concerned with, it is affected by biodiversity at the other levels.

Species diversity is the aspect of biodiversity with which we are most familiar because conservation legislation and treaties focus on preservation of species. In the first part of this exercise, we will focus on species diversity and various ways to quantify it. Diversity at the species level can be compared on a spatial scale by comparing alpha, beta, and gamma diversity.

Case Studies and Data

Explore these various indices of biodiversity and answer the questions that follow at home:

Alpha, Beta, and Gamma Diversity

Alpha diversity is the first-order diversity that exists within a habitat, i.e., the number of species found in some standardized area such as an acre, a square mile, or some naturally demarcated habitat area.

Beta diversity reflects the change (or turnover) in species composition over small distances, frequently between adjacent yet noticeably distinct habitat types. For example, beta diversity would be the difference between two different communities found on a mountain slope, such as lowland rain forest and montane evergreen forest. As habitat diversity increases, so does beta diversity.

Gamma diversity is the total species richness of a large geographic area, such as a continent. Gamma diversity combines the influence of alpha and beta diversity (Hunter, 1996).

$$\gamma = \alpha \times \beta. \tag{17.1}$$

Alpha and beta diversity values will differ depending on how you define each habitat. For example, you might consider ponds as a distinct habitat from surrounding meadow, or you might consider deep water, surface, shoreline, and meadow as distinct habitats. If we were interested in bird biodiversity, the two habitats would be relevant, but if we are interested in invertebrate biodiversity, the four-habitat subdivision would be relevant.

As an illustration of these categories, consider the following example: the tough dry rocky coastal algae A and B are found to coexist high above the high-tide line on the same small rocky island off Victoria, British Columbia. One species typically lives on the east-facing slopes and the other on the west-facing slopes. These two species of algae constitute the alpha species diversity on the vertical slopes of that small island. Now consider alga C—a different species that specializes on the flat tidal regions of the same rocky island (i.e., a different habitat). The average alpha diversity for algae on that island is 1.5 (the average of two high-altitude species and one tidal plain species). Because the gamma diversity for the island is three species, the beta diversity is 2 (3 divided by 1.5). If you also consider species D, which is found on islands farther south, your global gamma diversity for these algae would now be four species.

Jaccard's index gets at beta biodiversity as a ratio of common species to habitat specialists. Imagine finding ten species of arthropod in the pond and ten in the meadow, but two of the species can be found in both habitats. Because 16 species are specialists, Jaccard's index of beta diversity would be 2/16. This differs from the value you would get by simply dividing gamma diversity by alpha diversity (18/10).

Unfortunately, alpha diversity often increases as a result of human disturbance (e.g., road cuts). For example, higher alpha diversity in logged forests, compared to unlogged forests, might result from the edge effect created by road cuts. Therefore, we should be careful to look for the underlying reason for high alpha diversity when using it to inform conservation policy.

Let's consider some data on bird biodiversity on the islands Trinidad and St. Kitts collected by Ricklefs and Cox (1977). Each island had nine different habitats. Trinidad had a total of 108 species of birds with an average of 28.2 species/habitat, and St. Kitts had a total of 20 species with an average of 11.9 species/habitat.

Questions to Work on Individually Outside of Class

1. What are the alpha, beta, and gamma diversities for birds on both islands?
2. Which island is more diverse in bird species? Why?
3. What does the beta value in each case signify? What does a higher beta value indicate?

Dominance-Diversity Curves (Questions 4–10)

Sometimes, when comparing the diversity of different regions, it is helpful to examine dominance-diversity curves, which take into account the relative abundance of the different taxa. We can then look at individuals within species or species within higher taxa, depending on our data and the questions we are asking. A dominance-diversity curve plots the number of individuals or species (expressed as a proportion of the total number, or pi) as a function of that taxon's abundance rank (the highest rank is 1 and is assigned to the taxon that has the most individuals or species). A *flat* curve indicates that there is relatively low "dominance" of any taxon in that region. That is, there are many taxa in this region, but even the most common taxon accounts for a relatively small proportion of the total number. *Steep* curves indicate that the majority of a sample are members of the most common taxa and that the remaining taxa in the region are relatively rare.

4. Plot the diversity curves for regions 1 and 2 in table 17.1a. Plot Rank (x-axis) versus log(pi) values (y-axis) for each region. How do the slopes of the dominance diversity curves differ for the two regions?

TABLE 17.1a.
Illustrative diversity data for two regions

Taxa	No. of species	Rank	pi	log (pi)
		Region 1		
Mammals	52	2	0.1733	−0.7611
Birds	51	3	0.1700	−0.7696
Fish	53	1	0.1767	−0.7528
Insects	48	5	0.1600	−0.7959
Arachnids	49	4	0.1633	−0.7869
Mollusks	47	6	0.1567	−0.8050
Total	300			
		Region 2		
Mammals	19	5	0.0633	−1.1984
Birds	31	4	0.1033	−0.9858
Fish	50	3	0.1667	−0.7782
Insects	91	2	0.3033	−0.5181
Arachnids	10	6	0.0333	−1.4771
Mollusks	99	1	0.3300	−0.4815
Total	300			

5. Now complete table 17.1b by regrouping the data in table 17.1a. Plot the data on a second graph.

TABLE 17.1b.
Recalculated diversity comparison.

Taxa	No. of species	Rank	pi	log (pi)
		Region 1		
Vertebrates				
Arthropods				
Mollusks				
Total	300			
		Region 2		
Vertebrates				
Arthropods				
Mollusks				
Total	300			

6. How do the regions compare now?

7. What taxonomic level is mammal (i.e., *Mammalia*)? What taxonomic level is vertebrate (i.e., *Vertebrata*)? What taxonomic level is mollusk (i.e., *Mollusca*)?

8. Why would it be inappropriate to use mammals and mollusks in the same dominance-diversity plot? Why was it more appropriate to use vertebrates and mollusks in the same plot?

9. If all of the mollusks in these regions were bivalves, would your comparison in question 4 be a legitimate comparison of taxa?

10. If you had sufficient funds to support the conservation of just *one* of these regions, which one would you pick? Why?

Simpson and Shannon Indices (Questions 11–13)

Another important way to quantify diversity is to use the "Simpson Diversity Index." The Simpson index (D) takes into account the size of each population, and thus its relative contribution to the community.

$$D = 1\Big/\sum pi^2. \tag{17.2}$$

As above, "pi" is the proportion of abundance of a certain species relative to the total number of individuals in the sample. As pi values increase, D becomes smaller and the biodiversity is less evenly distributed.

Similar to the Simpson index, the Shannon index (H') combines information on species and their relative abundance into one number. The Shannon index was originally derived from information theory. Shannon was interested in how information was transmitted through phone lines, but his work has been widely applied (see chapter 9 on island biogeography and chapter 18 on environmental uncertainty). What does a high or low Shannon index tell us specifically? Suppose there exists a community and we randomly pick an individual organism out of it. If we are able to guess its identity beforehand (before picking it) either because one or two species in this community are very abundant, or because there are only a few species in this region, then the species diversity will be low and the value for H' will be low as well. H' is a measure of sampling uncertainty; in other words, how sure you can be, *a priori*, just what species will be in your sample. On the other hand, if we are not able to guess what the identity of that randomly picked individual is beforehand because the number of species in that community is high and the species are about equally abundant, then the H' value will be high, i.e., an indication of greater diversity. H', is actually a measure of sampling uncertainty.

$$H' = -\sum pi \times \ln(pi). \tag{17.3}$$

Note that we add a minus sign in front of the summation to make the value positive.

Compare the biodiversity of trees in one hectare of tropical forest in three Latin American countries by calculating Shannon and Simpson indices (tables 17.2, 17.3, and 17.4).

11. Calculate the Simpson and Shannon indices for forests in Bolivia, Panama, and Costa Rica.

12. Explain in words how the three biodiversities differ among the forests.

13. Which index of biodiversity is the most sensitive to the relative abundances of species?

TABLE 17.2.
Panama biodiversity.

Species	No. of individuals	pi	pi × pi	pi × ln(pi)
1	17			
2	17			
3	17			
4	17			
5	17			
6	17			
7	17			
8	17			
9	17			
10	17			
Total =	170		$\sum(pi)^2 =$	$-\sum pi \times ln(pi) =$

TABLE 17.3.
Costa Rica biodiversity.

Species	No. of individuals	pi	pi × pi	pi × ln(pi)
1	54			
2	11			
3	17			
4	9			
5	12			
6	17			
7	13			
8	4			
9	17			
10	16			
Total =	170		$\sum(pi)^2 =$	$-\sum pi \times ln(pi) =$

TABLE 17.4.
Bolivia biodiversity.

Species	No. of individuals	pi	pi $\times$ pi	pi $\times$ ln(pi)
1	10			
2	10			
3	10			
4	10			
5	10			
6	10			
7	10			
8	10			
9	10			
10	10			
11	10			
12	10			
13	10			
14	10			
15	10			
16	10			
17	10			
Total =	170		$\sum(\text{pi})^2 =$	$-\sum \text{pi} \times \ln(\text{pi}) =$

Small-Group/In-Class Exercise

Fighting for Biodiversity

You have demonstrated how biodiversity can be quantified with a variety of measures and that inferences from each measure differ. The most basic measure, alpha diversity, simply gives us the number of species (species richness) in a particular location. You have also seen how we can quantify the evenness or rareness of species, and the diversity due to habitat turnover in a location (beta diversity). However, as you demonstrated above, the taxonomic level at which you quantify biodiversity can greatly influence any comparison.

The biodiversity of a location may reflect high local endemism. An endemic species is one that is found only in a certain geographic area. An example is the bald eagle, which is endemic to North America. A local endemic is a species that is found only in a small area (such as one of the Galapagos Islands). Organisms can be endemic to a location for two different reasons: because they originated in that place and never dispersed, or because now their range has been reduced to only a small fraction of what it was initially (Brown, 1998).

Your job today will be to evaluate the biodiversity in one of four very different locations. The Bill and Melinda Foundation (subsequently referred to as the BMF) has budgeted 15 billion dollars "for the purchase and protection of the most important biodiverse real estate in the world." BMF researchers have compiled data on the four finalists, which they think they can afford to buy.

Your team will be assigned to one of these locations and must draft an argument why that particular location is the most important and why it deserves the protection that only the BMF can offer.

1. Use the tools you learned in part I of this exercise to make your case.

2. Remember that Bill is a hotshot CEO and will be most influenced by *quantitative* comparisons of your pet region with the other finalists.

3. Use all the different measures of biodiversity available.

4. Carefully examine the data available to the other teams and prepare questions and counterarguments for the case they are likely to make.

Missouri Missouri's biodiversity (tables 17.5 and 17.6) results from a combination of geographic location, physiographic diversity, and climatic history. Missouri contains parts of

TABLE 17.5.

Estimated numbers of plant species in various taxonomic groups and their status in Missouri.

Taxonomic group	Status							Comments
	Native	Introduced	Extirpated	Federal list*	State list*	Endemics	Relics	
Vascular	1,866	688	66	4, 5, 35	164, 66, 42	50	38	Most species, endemics, and relics in Ozarks. Greatest threats in Mississippi Lowlands.
Bryophytes (mosses and liverworts)	400	0	3	0	47, 18, 0	2	12	Approx. 2/3 of all MO species are found in the Ozarks. Many grow in restricted habitats sensitive to alteration.
Fungi	1,500 ?	?	?	0	0	?	?	Little taxonomic or census work done in Missouri
Algae	?	?	?	?	?	?	?	Little taxonomic or census work done in Missouri
Lichens	370 ?	?	?	0	13, 4, 2	?	?	There is much more to learn about this group. Many species confined to restricted habitats sensitive to alteration.

Source: The report of the Missouri Biodiversity Taskforce, 1992. *Endangered, rare, watchlist.

TABLE 17.6.
Estimated numbers of animal species in various taxonomic groups and their status in Missouri.

Taxonomic group	Native	Introduced	Extirpated	Federal list*	State list*	Endemics	Relics	Comments
Mammals	74	4	7	5, 0, 2	4, 7, 2	0	0	Extirpated and endangered species are generally large and wide ranging or require specialized habits.
Birds	173 breeding 147 winter 149 migrants	7	13	6, 0, 2	10, 12, 9	0	0	Greatest threats are habitat loss and fragmentation
Amphibians/ reptiles	108	0	1	0, 0, 5	6, 8, 17	6	6	MO has six Ozark endemic salamanders. Greatest threats in Mississippi Lowlands.
Fish	201	10	3	0, 4, 9	18, 20, 16	18	5	25% of N. American freshwater species occur in MO. High diversity of primitive and modern species. Most species and endemics in the Ozarks.
Aquatic crustaceans (decapods, amphipods, isopods)	64	0	2	0, 0, 8	3, 9, 9	26	0	Most species are endemics in Ozarks.
Mollusks (aquatic snails, mussels, bivalves)	129	1	1	4, 0, 7	22, 3, 0	11	2	Figure include aquatic snails, mussels and other bivalves only. Most species in Ozarks.
Aquatic insects	2,500 ?	?	?	0, 0, 6	3, 2, 3	15	15	Gross estimate only, little current information. Most species and endemics in Ozarks.
Terrestrial insects and spiders	15,000 ?	?	?	0, 0, 1	0, 0, 4	?	?	Gross estimate only, little current information.
Other (annelids, flatworms, nematodes, protozoans, etc.)	?	?	?	?	1, 0, 0	?	?	Very little is known about many groups of inconspicuous, mainly microscopic animals.

Source: The report of the Missouri Biodiversity Taskforce, 1992. *Endangered, rare, watchlist.

the Great Plains, eastern deciduous forest, northern boreal forest, desert southwest, and coastal plain habitats (figure 17.2). Specifically, Missouri contains 89 different terrestrial communities, including 26 different forest communities, 9 savannah, 14 prairie, 6 glade, 9 cliff, 3 talus, 2 stream bed, 15 wetland, and 5 cave. The Ozarks in southern Missouri are one of the oldest inhabitable land masses in North America. And that's only the terrestrial communities!

The two great North American rivers, the Mississippi and the Missouri, meet in Missouri, and their tributaries and seasonal floods contribute to the aquatic diversity in the state. The complex geographic structuring of the Ozarks also provides a multitude of habitats and microclimates. There are six endemic salamanders in the Ozarks, including the giant hellbender. The paddlefish is one of the oldest sturgeons (its closest relative is found only in the Yangtze River in China).

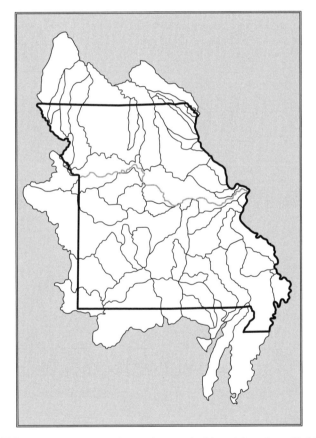

FIGURE 17.2. Missouri's geologic history has resulted in a rich variety of habitat types.

Madagascar An island long separated from the African mainland, Madagascar (figure 17.3) has one of the richest and most varied biota found in the world (Goodman, 1995). There is probably no other place on earth with such a concentration of unique plant and animal species (table 17.7). Of the 10,000 to 12,000 plant species found here, 80% are endemic (including more than 1000 species of orchids)!

FIGURE 17.3. Madagascar.

Topographically, Madagascar has a wide array of macrohabitats. The land on the east coast of the island rises to heights of 3000–5,000 feet behind a narrow coastal plain. A hilly plateau occupies the central part of the island. This plateau gradually slopes off to the west; here a few volcanic uplands are found (there is no current volcanic activity). The highest volcanic upland has an elevation of 9,450 feet. Precambian granitic rock, which underlies the eastern and central regions of Madagascar, has weathered into deep, red, and nutrient-riched soil. The central plateau is dotted by huge granite domes. Sedimentary formations from the Jurassic era are seen in the west.

TABLE 17.7.
Madagascar biodiversity.

	Species	Endemic
Mammals		
bat	28	10
tenrec	30	30
rodent	11	11
carnivore	8	8
Birds	198	106
Reptiles		
gecko	63	53
chameleon	54	54
crocodile	1	
sea turtle	4	
land tortoise	5	4
skink	48	47
girdle tailed lizard	12	11
snake	60	
Amphibians		
frogs	150	148
Plants		
total plant families = 191		8
total plant genera = 1,300		240
total plant species = 8,000		6,400
palms	175	
orchids	1,000	

Water-laden air from the Indian Ocean on the east has resulted in rainforest on the east side of the island. In the west and northwest, there is dry tropical forest, and in the south and southwest, there is spiny desert. Since land clearing (by people, fire, and cattle) has taken place principally in the central region, the remaining natural areas (15%) are found only around the perimeter of the island. Only 2% of the island is protected in parks and reserves.

Eastern Mongolia Mongolia has many ecosystems that are critical for regional biodiversity. These include wetlands, dryland, steppe, and temperate forests (figure 17.4). Notable among these is the Eastern Steppe, which is probably the world's largest and last remaining unspoiled great plains ecosystems. It is gifted with astonishingly rich biodiversity (table 17.8). Eastern Mongolia contains many endemic plant species; it is home to the white-napped crane, a rare endemic bird, and herds of the very rare Mongolian gazelle.

Eleven percent of the total surface water runoff in Mongolia is caused by the two important rivers in the region: the Khalkh and the Kherlen. The mean annual precipitation ranges between 210 and 397 mm. Mining is an important industry in Mongolia. Fluorite spar, oil, tin, uranium, gold, and silver are some of the largest mineral deposits. Coal mining is also crucial as an energy source and for the economy. Dust and water effluent discharge from these mines results in great damage to the rivers and streams, and also results in soil erosion. Lack of land restoration following mining is also a huge problem.

FIGURE 17.4. Eastern Mongolia.

TABLE 17.8.
Eastern Mongolian fauna.

	Species	Very rare species	Species listed in CITES
Mammals			13-mammals
Insectivores	6	1	
Chiroptera	7	1	
Lagomorpha	4		
Rodentia	32	1	
Carnivora	14	1	
Artiodactyla	8	5	
Birds	327	21	54-birds
Amphibians			2-amphibians
Caudata	1	1	
Anura	3	1	
Reptiles			4-reptiles
Lacertilia	3		
Serpents	5		
Fish	45	3	2-fishes
Insects	5874	6	

Notes: CITES is the convention on the international trade in endangered species. If a species is listed, international trade in living or dead members of that species is severely restricted. This is enforced by customs officers of the signatory nations.

Another potential pressure on the ecosystem comes from livestock grazing, which has been shown to decrease alpha diversity in nearby China. Overall, there are clear trends of decreasing populations of many species in all parts of Mongolia, including the eastern Steppe. In southern Mongolia, the Gobi National Conservation Park has been successful in maintaining healthy populations of many species of vertebrates which are not faring well in other parts of the country, including the Mongolian gazelle; however, no such park has yet been established in the eastern Steppe.

Note: CITES is the convention on the international trade in endangered species. If a species is listed, international trade in living individuals or non-living samples (or even parts such as feathers, fur, or flowers) is severely restricted. CITES is enforced by customs officers of the signatory nations.

Galapagos

Until the discovery of the Galapagos Islands in 1535 by the Bishop of Panama, Fray Tomas de Berlanga, the flora and fauna of this archipelago evolved in isolation, producing marvelous species. These islands were created by volcanic activity on the Pacific Ocean floor only 3–4 million years ago. In the Galapagos archipelago, there are 13 large islands, 6 small islands, about 40 islets, and a multitude of unnamed small rocks (figure 17.5). The nearest landmass, Ecuador, is 960 km away to the east. The climate is strongly influenced by the conjunction of contrasting oceanic currents: the relatively cold Humboldt Current from the Antarctic, and warm tropical waters from the Gulf of Panama. These

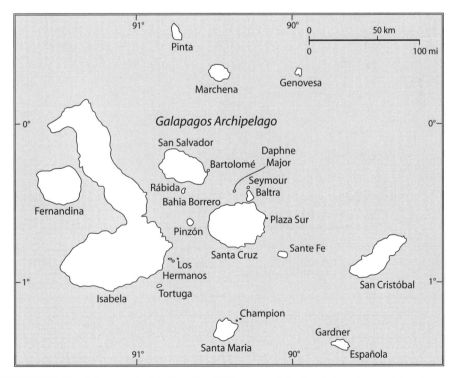

FIGURE 17.5. The Galapagos Archipelago lies 960 km west of the South American mainland.

shifting currents cause two distinct seasons, a dry, cool season and a shorter, hot season, periodically marked by heavy rainfall. In addition to open sea and rocky islets, the archipelago has six distinct vegetation zones, defined by altitude and rainfall patterns. These include the shore zone, arid zone, transition zone, *Scalezia* zone, *Miconia* zone, and pampa zone.

The Galapagos Islands have roughly 625 native taxa (species, subspecies, and varieties), and 190 introduced species (tables 17.9–17.11). Two hundred and twenty-five of these species are endemics. Introduced plants and animals have had a heavy impact on the landscape and native fauna and flora of the Galapagos Islands. Some of the worst problems are caused by domestic animals such as goats and their parasites. In addition, uncontrolled fishing is a major threat to marine biodiversity.

The Galapagos Islands hold significant cultural and scientific heritage. In 1835, Charles Darwin visited the Galapagos while on his voyage aboard the *HMS Beagle*. His observations on species diversity between the islands were later used to develop the theory of natural selection. These volcanic islands have been called a unique living museum and showcase of evolution.

TABLE 17.9.
Herps of the Galapagos.

Type	Species recorded	Indigenous residents	Marine migrants	Introduced	Endemic species
		Status			
Tortoise	1	1			1
Turtles	4	1	3		
Iguanas	3	3			
Lava lizards	7	7			7
Geckos	9 (10?)	6 (7?)		3	5 (6?)
Snakes	3	3			3
Sea snakes	1		1		

TABLE 17.10.
Mammals of the Galapagos.

Type	Species recorded	Residents	Migrants	Endemic species	Endemic subspecies
		Status			
Sea lions	2	2		1	1
Rodents	4	4			
Bats	2	2		1	1
Whales and dolphins	25		25		

TABLE 17.11.
Birds of the Galapagos.

Type	Species recorded	Status			Endemic species	Endemic subspecies
		Residents	Migrants	Vagrants		
Seabirds	47	19	4	24	5	8
Waterbirds	22	11	1	10	2	3
Shorebirds	34	2	16	16		1
Diurnal raptors	3	1	2		1	
Nightbirds	3	2		1		2
Larger landbirds	8	5	1	2	1	
Aerial feeders	6	1	1	4	1	
Smaller landbirds	29	20		9	18	2

References

Brown, J. H., and M. V. Lomolino, 1998. *Biogeography*. Sunderland, Mass.: Sinauer Associates..

Castri, F., and T. Younes, 1996. *Biodiversity, Science and Developmen: Towards a New Partnership*. Paris: CAB International.

Goodman, S. M., and B. D. Patterson, 1995. *Natural Change and Human Impact in Madagascar*. London: Smithsonian.

Hunter, M. L., Jr. 1996. *Fundamentals of Conservation Biology*. Cambridge, Mass: Blackwell.

Jackson, M. H. 1993. *Galapagos*. Calgary: Calgary University Press.

Ricklefs, R. E., and G. W. Cox, 1977. Morphological similarity and ecological overlap among passerine birds on St. Kitts, British West Indies. *Oikos* 29: 60–66.

Swash, A., and R. Still, 2000. *Birds, Mammals, and Reptiles of the Galapagos Islands*. London: Yale University Press.

UNEP. 1992. *Convention on Biological Diversity, June 1992*. Nairobi: United Nations Environment Programme.

WRI, IUCN, UNEP. 1992. *Global Biodiversity Strategy: Guidelines for Action to Save, Study and Use Earth's Biotic Wealth Sustainably and Equitably*. Baltimore: World Resources Institute Publications.

18 Environmental Predictability and Life History

Bobbi S. Low and Stanton Braude

Introduction and Background

Global climate change is leading to serious problems for innumerable species. In many cases, it is because the extremeness and the predictability of important environmental conditions are being changed by human activity. The Intergovernmental Panel on Climate Change (2002) reviewed 2,500 scientific papers on changes in timing of seasonal events in plants and animals. Eighty per cent of the 500 species studied had experienced shifts in their ranges, or the timing of migration, reproduction, or growing season, as a result of climate shifts. Sometimes these shifts have led to serious mismatches between the cues organisms use to "predict" changes, and the conditions (e.g., food availability) that follow.

We know that environmental conditions shape the life histories and behavior of all organisms (e.g., chapter 1). Before you go on, think for a few minutes about the ways in which an environment might be predictable. What does it mean to be unpredictable?

If an environment never varies, it is obviously predictable. If the environment does vary, two things make a difference in how the changes affect an organism's life history (and thus how we can manage it): how much do things vary (the range of variation) and how predictable is the timing of changes? These parameters influence an organism's life history (chapters 4 and 6) and lead to questions such as: When is it best to lay eggs or get pregnant? When is it best to forage, or to hunt? When, if ever, does it make sense to hibernate or aestivate?

Think for a moment of three frogs. One lives in a warm, wet, constant-temperature forest in Indonesia; one lives in southern Ontario, where rainfall is seasonal and there are cold winters; and one lives in the center of Australia, where heat and dryness are extreme, and rainfall (which is rare) is not more likely to come in one season than another. Their lives will be extremely different. In this chapter, we will explore ways to figure out just how their lives will differ. Though their influences are separate, extremeness and predictability both influence life histories.

Think now of the second two frogs above: one in Ontario, one in central Australia. Organisms can respond to extreme environmental fluctuations in a variety of ways: they can avoid or tolerate extremes, or (if conditions favor this, and the genetic variation exists) they can develop specializations to cope. If fluctuations are predictable in time (e.g., seasonal), organisms can "anticipate" them. You will not be surprised to learn that frogs in Ontario hibernate during the cold winters (there is clearly a better and a worse time for reproduction). And in many dramatic examples of seasonal migration, for example, the organism begins migrating before there is any significant change in local conditions. Thus, hummingbirds and wildebeest start migrating away from environments that will shortly deteriorate *before* the deterioration actually begins.

The ability to anticipate environmental change, because the change is regular in its occurrence, allows organisms to optimize the timing of important life history events such as flowering, seasonal migration, mating and parturition. Basically, organisms are likely not to wait for the unfavorable conditions to occur, but to find some correlated environmental trait on which to cue. Easily detected environmental traits are the ones most likely to be used. So hummingbirds and frogs in northern North America do not wait until it is cold and food disappears to hibernate or leave; they use the "cheap cue" of shorter day length (temperature itself varies widely in this region in late summer and early fall, and is not a reliable cue). Organisms use cues to get reliable information from the environment: if short day length, for example, signals a coming environmental deterioration, an organism able to discern the shift in day length gets information about what will happen. On the other hand, if an environmental factor is unpredictably timed, we expect the evolution of alternative strategies to tolerate extremes or to change investment in life effort rapidly. Unpredictability imposes a cost.

Return to the three frogs. If we make a matrix of the season and water availability over nine years, not only can we see dramatic differences that tell us about predictability, but we can devise a simple way to figure out how predictable the environment is—either because it is constant, or because it changes regularly. Frog 1 (in Indonesia) lives in a water regime like table 18.1a (we are using three, rather than four, seasons for simplicity). For the Indonesian frog, with regard to water availability, season doesn't matter: any time is as good as any other for breeding. How predictable is water availability in this environment?

Frog 2, from Ontario, faces some cold and dry periods, and some warm and dry periods—and one season that always has plenty of water (table 18.1b). In our data, clearly there is one best season for laying eggs.

Now, Frog 3 (table 18.1c): Central Australia, as you'll learn in more detail below, has the least predictable rainfall of all the locations studied in this way. There isn't much rain (extremeness), but, equally important from a life history perspective, rain is about equally likely to fall in any amount in any season. This is why the matrix is all "3's." In nine years, each level of rainfall fell three times in each of the three seasons; our prediction is that it will continue to be equally likely in each season. This presents the frog with a real dilemma! It literally cannot predict anything ahead of time (which might save effort) about rainfall and water availability. But some other strategies (besides "wait and see") can be important. The Australian frog, *Cyclorana*, burrows and aestivates. The depth of its burrow matters. Too shallow, and the frog gets wet and responds to rains too light to make feeding and breeding possible. Burrowing too deep means the frog never gets wet again. But at the right depths the frog's environment has become predictable: *If* it gets wet, *then* it will have enough water to feed and breed. Its behavior has created a threshold, making its cue (getting wet) predictable.

You have probably begun to have an intuition about these matrices. If things are always the same, they are predictable. Further, if things never change, you can ignore that variable, because it gives you no information about better-versus-worse times to feed or reproduce, for example. But you also see that "predictability" is not singular: things can be predictable (as for the Ontario frog in table 18.1b) because they fluctuate regularly or because they are constant—and these mean very different things for the animal's life. Look back at the three matrices: two (tables 18.1a and18b) are completely predictable, but the matrices look

TABLE 18.1.

Occurrences of rainfall in different seasons over a period of nine years in (a) an Indonesian rainforest, (b) southern Ontario, and (c) central Australia.

(a) Indonesian Rainforest			
	Spring	Summer	Fall/Winter
Very dry	0 years	0	0
Some water	0	0	0
Plenty of water	9	9	9

(b) Southern Ontario			
	Spring	Summer	Fall/Winter
Very dry	0 years	0	9
Some water	0	9	0
Plenty of water	9	0	0

(c) Central Australia			
	Spring	Summer	Fall/Winter
Very dry	3 years	3	3
Some water	3	3	3
Plenty of water	3	3	3

different and the frogs' life histories will differ. The third (Australian) frog lives in a costly, unpredictable environment with regard to the water regime.

This simple approach is a modification of Colwell's (1974) adaptation of Shannon and Weaver's (1949) information theory. Shannon, an engineer, was originally seeking a way to measure information (just what an organism tries to get from the environment!). The other descendant of Shannon-Weaver information theory you encountered in an earlier chapter is the species diversity index. That uses some of the same calculations we use here. When species diversity is high, it translates to mean "sampling uncertainty is high."

These approaches allow us to separate predictability into (1) predictability based on *contingency* (seasonality, which the Ontario frog experiences, and which for an engineer is information), and (2) predictability based on *constancy* (true for the Indonesian frog; in Shannon-Weaver terms, this is noise). This is important because the two different kinds of predictability, as we noted, have quite different selective impacts. Living in a seasonal environment poses quite different problems compared to living in a constant one!

A nice feature of this "quick and dirty" approach is that the measures are all transformed so that they cannot be less than zero or greater than one, and that predictability (P) is the sum of constancy (C) and contingency (M) (i.e., $P=C+M$). We used three intuitively clear examples: Indonesia: $P=1$ (perfectly predictable), $C=1$ (completely constant), and $M=0$ (there is no rainfall seasonality or contingency at all); Ontario: $P=1$, $M=1$ (completely contingent on season), and $C=0$; and Australia: $P=0$, $M=0$, and $C=0$ (nothing whatsoever is predictable about rainfall). Of course, we never see these "perfect" patterns in nature, but we can look for biases in the distribution of numbers in the matrix (more below).

TABLE 18.2.
Construction of Colwell's predictability, constancy and contingency matrix.

States (i)	Time 1	Time 2	Time 3	t = number of times
		Time (j)		
State 1	N_{ij} or N_{11}	N_{12}	N_{13}	$Y_1 = \Sigma N_{ij} =$ $N_{11} + N_{12} + N_{13}$
State 2	N_{21}	N_{22}	N_{23}	$Y_2 = \Sigma N_{ij} =$ $N_{21} + N_{22} + N_{23}$
State 3	N_{31}	N_{32}	N_{33}	$Y_3 = \Sigma N_{ij} =$ $N_{31} + N_{32} + N_{33}$
s = the number of states	$X_1 = \Sigma N_{ij} =$ $N_{11} + N_{21} + N_{31}$	$X_2 = \Sigma N_{ij} =$ $N_{12} + N_{22} + N_{32}$	$X_3 = \Sigma N_{ij} =$ $N_{13} + N_{23} + N_{33}$	$Z = Y_1 + Y_2 + Y_3$ $= X_1 + X_2 + X_3$

This method can be used to report the full spectrum of predictability. To calculate P, C, and M (predictability, constancy, and contingency), we create a matrix of time and state combinations, just as in table 18.1. Each cell is filled with a tally of the number of observations for which each particular state has occurred during a particular time. Times can be seasons of the year, hours of the day, or phases of the moon, for example, but seasonality is most often used. Examples of environmental states that vary include climatic categories, or the vegetative states of a resource (e.g., flowering, fruiting, etc.).

There are two sets of patterns you will look for. First, there are some patterns in the matrix itself—certain distributions allow you to say "C is high" (table 18.1a) or "M is high" (table 18.1b). Second, once you calculate P, C, and M from the matrix, you can connect the values of P, C, and M to the life histories of organisms. Low P has high cost and clear selective pressures. High P means nothing by itself, but high/low C and high/low M have clear selective importance.

With this approach you can look for patterns in the matrix that will tell you about life history even before you go through the calculations (which will work even when a pattern is not clear to you). So, the Indonesian frog faced a constant regime: no matter the season, the water availability was in the same category. You would describe that matrix as "having only one nonzero row," reflecting constant conditions (table 18.1a). Simple—and you can say something about life history from it. The Ontario frog's matrix has "only one nonzero value in each row and column"—a perfectly contingent matrix (if you look in column 1, the only nonzero value will be in row 3, for example). And the Australian frog, which experiences zero predictability (any state is equally likely in any season) has a "homogeneous matrix."

In this exercise, you will seek two kinds of bottom lines: matrix patterns, and life history effects. You have already done so for the three frogs. Now we will learn the full procedure (table 18.2).

Once your data on "the number of times a state has been observed in a particular time period" have been entered in the cells of the matrix, you can total the number of times a particular state has been observed overall (Y_1, Y_2, etc. in table 18.2), and the number of observations in each time period (X_1, X_2, etc. in table 18.2). To determine predictability, constancy, and contingency we must next calculate some intermediate variables: uncertainty with regard to time ($H(x)$), uncertainty with regard to state ($H(y)$), uncertainty with regard to the interaction between state and time ($H(xy)$), and uncertainty with regard to state when time is known ($Hx(y)$).

$$\text{Uncertainty with regard to time: } H(x) = -\sum_{j=1}^{t} \frac{X_j}{Z} \log \frac{X_j}{Z}, \tag{18.1}$$

$$\text{Uncertainty with regard to state: } H(y) = -\sum_{i=1}^{s} \frac{y_i}{Z} \log \frac{y_i}{Z}, \tag{18.2}$$

$$\text{Uncertainty with regard to interaction: } H(xy) = -\sum i \sum j \frac{N_{ij}}{Z} \log \frac{N_{ij}}{Z}, \tag{18.3}$$

$$\text{Uncertainty with regard to state (if time is known): } Hx(y) = H_{(xy)} - H_{(x)}. \tag{18.4}$$

Predictability, constancy, and contingency can then be calculated because

$$\text{Predictability } P = 1 - \frac{Hx(y)}{\log s}, \tag{18.5}$$

$$\text{Constancy } C = 1 - \frac{H(y)}{\log s}, \tag{18.6}$$

$$\text{Contingency } M = \frac{H(x) + H(y) - H(xy)}{\log s}. \tag{18.7}$$

Although the calculation of these intermediate variables can feel tedious, it is important. As you proceed through this exercise, be sure you look at each matrix you analyze, and try to predict from the pattern the approximate P, C, and M (just aim for "high," "low," and "I can't tell yet").

Bottom Lines: Why We Do This

Certain matrix patterns make clear life history predictions. These are important in understanding how predictability, constancy and contingency affect life histories. It may help if you draw a specific matrix for each case:

When the matrix is homogeneous, we cannot predict the state (any state is equally likely in any season; $P=0$). *This is an "expensive" environment in which an organism faces changes in environmental states, but cannot use a cheap cue to be ready.*

When only one state is found (i.e. there is only one nonzero row total), constancy is maximal. *In this case (as in the Indonesian frog above) the variable gives no better or worse information about when to initiate important activities like breeding. Other parameters will determine, e.g., optimal timing for breeding.*

When there is only one non-zero cell in each combination of row and column, contingency is maximal. *In this case, the organism will probably evolve to use some cue that also is contingent (e.g., day length, which predicts temperature roughly in temperate environments), especially if it is easily perceived ("cheap"). The Ontario frog breeds seasonally.*

In the next "matrix bottom line" what you can predict about life history depends on additional factors; here, you'll need to look again at the matrix:

When each state is found in equal frequency (i.e., the row totals are equal), constancy is minimal. *If M is also low, the total predictability will be low. If M is high, the matrix may be seasonal (above) and the organism will cue.*

(When there are equal observations in each time period [i.e. the column totals are equal], there is no seasonal bias in our sampling; the intermediate variable "uncertainty with regard to season" is maximal. This does not mean much about bottom lines: all the examples had equal numbers of observations in each season!)

TABLE 18.3.
Sample matrix.

States	Time blocks			
	Early	Middle	Late	
Warm	5	2	0	$Y_1 = 7$
Cold	0	3	5	$Y_2 = 8$
$s = 2$	$X_1 = 5$	$X_2 = 5$	$X_3 = 5$	$Z = 15$

Example

Let us consider a simplified example in which we have three *seasons* (time blocks) and two different environmental conditions or *states* (table 18.3). The seasons can be early (January–April), middle (May–August), and late (September–December). The states in this simplified example are warm and cold. We can now create a matrix representing the possible states in these seasons. Here, the numbers in each cell (N_{ij}) are simply the number of observations that were in the indicated state, in the indicated season (e.g., of the 15 total observations, five were "early" and "warm"; $N_{11}=5$).

$$H(x) = -\sum_{j=1}^{t} \frac{X_j}{Z} \log\frac{X_j}{Z} = -\left[\left(\frac{5}{15}\right)\left(\log\left(\frac{5}{15}\right)\right)+\left(\frac{5}{15}\right)\left(\log\left(\frac{5}{15}\right)\right)+\left(\frac{5}{15}\right)\left(\log\left(\frac{5}{15}\right)\right)\right] \qquad (18.8)$$

$$= -[-0.159+-0.159+-0.159]=0.477,$$

$$H(y) = -\sum_{i=1}^{S} \frac{y_i}{Z} \log\frac{y_i}{Z} = -\left[\left(\frac{7}{15}\right)\left(\log\left(\frac{7}{15}\right)\right)+\left(\frac{8}{15}\right)\left(\log\left(\frac{8}{15}\right)\right)\right]$$

$$= -[-0.154+-0.146]=0.299$$

$$H(xy) = -\sum_i \sum_j \frac{N_{ij}}{Z}\log\frac{N_{ij}}{Z} = \left[\left(\frac{5}{15}\right)\left(\log\left(\frac{5}{15}\right)\right)+\left(\frac{2}{15}\right)\left(\log\left(\frac{2}{15}\right)\right)+\left(\frac{0}{15}\right)\left(\log\left(\frac{0}{15}\right)\right)+ \qquad (18.9)$$

$$\left(\frac{3}{15}\right)\left(\log\left(\frac{3}{15}\right)\right)+\left(\frac{0}{15}\right)\left(\log\left(\frac{0}{15}\right)\right)+\left(\frac{5}{15}\right)\left(\log\left(\frac{5}{15}\right)\right)\right]$$

$$= -[-0.159+-0.117+0+0+-0.139+-0.159]=0.574$$

$$Hx(y) = H(xy) - H(x) = 0.574 - 0.477 = 0.097 \qquad (18.10)$$

$$P = 1 - \frac{Hx(y)}{\log s} = 1 - \left(\frac{0.097}{0.3}\right) = 0.68 \qquad (18.11)$$

$$C = 1 - \frac{Hx(y)}{\log s} = 1 - \left(\frac{0.299}{0.3}\right) = 0.003 \qquad (18.12)$$

$$M = \frac{H(x) + H(y) - H(xy)}{\log s} = \frac{0.477 + 0.299 - 0.574}{0.3} = 0.673 \qquad (18.13)$$

Objectives of This Exercise

The purpose of this exercise is to understand how constancy and contingency affect environmental predictability and how they reflect important differences between locations or habitats. You will also begin testing hypotheses about the relationship between these environmental parameters and life history characteristics.

Case Study and Data

Much of our understanding of demographic and population genetic processes in fragmented wild populations comes from detailed studies of the imaginary yet charismatic desert rodent, the woggle. Woggles breed erratically in the spring, summer, and fall seasons, but successful rearing of young appears to depend on availability of the fruit of the plant *Datura albicans.*

Populations of *D. albicans* have been found throughout California, but its vegetative and reproductive cycles are strongly influenced by abiotic ecological factors such as climate and soil chemistry. On any given day you can find the flowers and fruit of the plant somewhere in the state. However, in any given location, *Datura* may be in any of its three life stages: growth, flowering, or fruiting.

It has been hypothesized that: "the reproductive success of woggles in wild populations results from the woggles' ability to respond to environmental cues that predict the vegetative state of *Datura.*" Fortunately, the Herbarium of the California Natural History Museum has a large collection of *Datura albicans* plants from the past 9 years. The exact date of collection is clearly printed on all of the pressed specimens. Thus we were able to generate predictability matrices for the vegetative state of *Datura* in six California deserts (tables 18.4a–18.4f). (For example: there were specimens collected in the spring of all 9 years in the Bakersfield desert and all of these specimens were in the growth phase; there were specimens in the collection from Berkeley in the spring in all 9 years and in one of those years the specimens were in the growth phase, in two of the years specimens were in the flowering phase, and in six of the years the specimens were in the fruiting phase.) Before you go on, identify which sites look like each of the frog matrices above!

The worksheets on the following pages are provided to help you calculate predictability, constancy, and contingency for the states of *Datura* in these six deserts (table 18.5).

TABLE 18.4.
Datura state matrices for six California locations.

(a) Bakersfield	Spr	Sumr	Fall		(b) Mojave	Spr	Sumr	Fall		(c) Berkeley	Spr	Sumr	Fall
Grow	9	9	9		Grow	9	0	0		Grow	1	1	1
Flowr	0	0	0		Flowr	0	0	9		Flowr	2	2	2
Fruit	0	0	0		Fruit	0	9	0		Fruit	6	6	6

(d) Death Valley	Spr	Sumr	Fall		(e) Lucerne	Spr	Sumr	Fall		(f) Alturas	Spr	Sumr	Fall
Grow	5	1	3		Grow	6	0	3		Grow	3	3	3
Flowr	3	3	3		Flowr	3	0	6		Flowr	3	3	3
Fruit	1	5	3		Fruit	0	9	0		Fruit	3	3	3

TABLE 18.5.
Worksheet for calculating components of *Datura* predictability.

		Bakersfield	Mojave

Three seasons

(Row = i)	N			Y_i
(Col = j)				Y_1
State				Y_2
				Y_3
	X_j	X_1	X_2	X_3	Z

Bakersfield Y_i

9	9	9	27	Y_1
0	0	0	0	Y_2
0	0	0	0	Y_3
X_j 9	9	9	27	

Mojave Y_i

9	0	0	9	Y_1
0	0	9	9	Y_2
0	9	0	9	Y_3
X_j 9	9	9	27	

$$X_j = \sum_{i=1}^{s} N_{ij} \quad \text{(column total)}$$

$$Y_j = \sum_{j=1}^{T} N_{ij} \quad \text{(row total)}$$

$$Z = \sum i \sum j N_{ij} \quad \text{(grand total)}$$

S = # of states

Bakersfield:
$X_1 = 9 \quad Y_1 = 27$
$X_2 = 9 \quad Y_2 = 0$
$X_3 = 9 \quad Y_3 = 0$
$Z = 27$

Mojave:
$X_1 = \quad Y_1 =$
$X_2 = \quad Y_2 =$
$X_3 = \quad Y_3 =$
$Z =$

*Uncertainty wrt time
$$H_{(x)} = -\sum_{j=1}^{t} \frac{X_j}{Z} \log \frac{X_j}{Z}$$

$$H_{(x)} = -3\left(\frac{9}{27} \log \frac{9}{27}\right) = .48 \qquad H_{(x)} =$$

*Uncertainty wrt state
$$H_{(y)} = -\sum_{i=1}^{s} \frac{y_i}{Z} \log \frac{y_i}{Z}$$

$$H_{(y)} = \left(\frac{27}{27} \log \frac{27}{27}\right) = 0 \qquad H_{(y)} =$$

*Uncertainty wrt interaction
$$H_{(xy)} = -\sum i \sum j \frac{N_{ij}}{Z} \log \frac{N_{ij}}{Z}$$

$$H_{(xy)} = -3\left(\frac{9}{27} \log \frac{9}{27}\right) = .48 \qquad H_{(xy)} =$$

*Uncertainty wrt state, time unknown
$$H_{x(y)} = H_{(xy)} - H_{(x)}$$

$$H_{x(y)} = .48 - .48 = 0 \qquad H_{x(y)} =$$

Predictability
$$P = 1 - \frac{H_{x(y)}}{\log s}$$

$$P = 1 - \frac{0}{.48} = 1 \qquad P =$$

Constancy
$$C = 1 - \frac{H_{(y)}}{\log s}$$

$$C = 1 - \frac{0}{.48} = 1 \qquad C =$$

Contingency
$$M = \frac{H_{(x)} + H_{(y)} - H_{(xy)}}{\log s}$$

$$M = \frac{.48 + 0 - .48}{.48} = 0 \qquad M =$$

P is minimum: homogeneous matrix

C is minimum: row totals are equal

M is minimum: columns identical

M is maximum: only 1 non-zero value in *each row* and *column*

C is maximum: only 1 non zero row

TABLE 18.5.
(*continued*)

	Berkeley				Death Valley				Lucerne				Alturas		

			Y_i				Y_i				Y_i				Y_i
1	1	1	3	5	1	3	9	6	0	3	9	3	3	3	9
2	2	2	6	3	3	3	9	3	0	6	9	3	3	3	9
6	6	6	18	1	5	3	9	0	9	0	9	3	3	3	9

X_j 9 9 9 27 X_j 9 9 9 27 X_j 9 9 9 27 X_j 9 9 9 27

Berkeley	Death Valley	Lucerne	Alturas
$X_1 =$ $Y_1 =$	$X_1 =$ $Y_1 =$	$X_1 =$ $Y_1 =$	$X_1 =$ $Y_1 =$
$X_2 =$ $Y_2 =$	$X_2 =$ $Y_2 =$	$X_2 =$ $Y_2 =$	$X_2 =$ $Y_2 =$
$X_3 =$ $Y_3 =$	$X_3 =$ $Y_3 =$	$X_3 =$ $Y_3 =$	$X_3 =$ $Y_3 =$
$Z =$	$Z =$	$Z =$	$Z =$

$H_{(x)} =$ $H_{(x)} =$ $H_{(x)} =$ $H_{(x)} =$

$H_{(y)} =$ $H_{(y)} =$ $H_{(y)} =$ $H_{(y)} =$

$H_{(xy)} =$ $H_{(xy)} =$ $H_{(xy)} =$ $H_{(xy)} =$

$H_{x(y)} =$ $H_{x(y)} =$ $H_{x(y)} =$ $H_{x(y)} =$

$P =$ $P =$ $P =$ $P =$

$C =$ $C =$ $C =$ $C =$

$M =$ $M =$ $M =$ $M =$

However, before you go on, examine these matrices, and jot down your predictions about whether *P*, *C*, and *M* will be high or low (or, you can't tell).

Questions to Work on Individually Outside of Class

1. Analyze the data above, and calculate *P*, *C*, and *M* (the predictability, constancy, and contingency) of the vegetative state of *Datura albicans* for each of the populations. This may seem like a lot of number crunching, but look at the six matrices carefully. Three of them have very simple patterns that match the frog example worked above. (Work at least one of the "obvious" examples to learn how the number crunching works. You certainly may work through the calculations for all six matrices, but you may only need to crunch the numbers for four of them.)

Bakersfield	*P*=	*C*=	*M*=
Mojave	*P*=	*C*=	*M*=
Berkeley	*P*=	*C*=	*M*=
Death Valley	*P*=	*C*=	*M*=
Lucerne	*P*=	*C*=	*M*=
Alturas	*P*=	*C*=	*M*=

2. Explain *why* the vegetative state of *Datura albicans* is, or is not, predictable in each case (and if predictable, is predictability due to constancy or contingency?).

Bakersfield
Mojave
Berkeley
Death Valley
Lucerne
Alturas

3. How do these data relate to the hypothesis that the reproductive success of woggles in wild populations results from the woggles' ability to respond to environmental cues that predict the vegetative state of *Datura albicans*? What other information would you need to evaluate this hypothesis fully?

Bring your clearly written answers to questions 1, 2, and 3 to class along with the worksheets provided for solving the matrices. Put your name on these pages and turn them in during class.

Small-Group/In-Class Exercise

Response to Constancy, Contingency, and Unpredictability

4. Thank you for analyzing the *Datura* data from the herbarium. Before continuing, please compare your results with those of other members of the group. Your calculated results should all agree. If there are differences, go back now to your actual calculations to find out where someone misunderstood or miscalculated.

5. Why is uncertainty with regard to time, $H(x)$, the same in all six populations? Why is uncertainty with regard to state, $H(y)$, the same in all populations except Bakersfield and

Berkeley? Why is there absolutely no uncertainty with regard to state once time is known, $Hx(y)$, in Bakersfield and Mojave?

6. What climatic factors could contribute to the fact that the vegetative state of *Datura* is constant in some of the populations but not in others? What climatic factors could contribute to the fact that the vegetative state of *Datura* is predictably contingent in some of the populations but not in others? You may want to consider comparisons of extremely different locations.

7. While you've been working on the *Datura* data set, we have been busy pulling together data on woggle survivorship and density in the six parks and reserves where woggles are still found (table 18.6). Your next job is to put our new woggle data together with your *Datura*

TABLE 18.6.
Further demographic data for six California woggle populations.

Desert park or reserve	Percent survivorship of young to maturity	Population density (per square km)	Approximate area of the park or reserve sampled (in square km)	Abundance
Bakersfield	0.01	0.2	10	
Mojave	0.58	1.2	6500	
Berkeley	0.30	11.0	1400	
Death Valley	0.20	3.0	2400	
Lucerne	0.53	18.0	1440	
Alturas	0.21	12.0	2620	

TABLE 18.7.
Predictability (P), constancy (C), and contingency (M) of rainfall for 11 deserts

Desert type	Years data	P	C	M
Subtropical				
Alice Springs, Australia	86	.111	.069	.042
Baghdad, Iraq	29	.571	.170	.401
El Centro, Calif., USA	28	.465	.328	.136
Aoulef, Algeria	29	.815	.755	.060
Cool coastal:				
O'Okiep, South Africa	21	.259	.041	.218
Arica, Chile	20	1.000	1.000	.000
Onslow, N.C., USA	74	.379	.199	.180
Rain shadow:				
San Juan, Argentina	84	.257	.202	.055
Phoenix, Ariz., USA	64	.184	.037	.147
Polar:				
Verkhoyansk, USSR	34	.838	.778	.060
Fort Good Hope, Canada	51	.904	.868	.036

data to test the hypothesis that the reproductive success of woggles in wild populations results from the woggle's ability to predict the vegetative state of *Datura albicans*. In addition to any arguments you were able to make about the *possibility* of woggles predicting *Datura* availability, please analyze these data in combination with the Herbarium data to give us a more complete evaluation. Graph paper may be provided to help you visualize correlations; use it. Be sure to use the appropriate graph or graphs to illustrate your conclusions.

8. What other data would help you test the hypothesis that the reproductive success of woggles in wild populations results from the woggles' ability to predict the vegetative state of *Datura albicans*? How could you get these data?

9. What life history traits in plants or animals would you predict might be associated with low predictability of rainfall in deserts? In addition to table 18.7 from Low (1978), what type of data could you collect to test that prediction?

References

Colwell, R. K. 1974. Predictability, Constancy, and Contingency of Periodic Phenomena. *Ecology* 55(5):1148–1153.

Felsenstein, J. 1985. Phylogenies and the comparative method. *American Naturalist.* 125(11): 1–15.

Low, B. S. 1978. Environmental uncertainty and the parental strategies of marsupials and placentals. *American Naturalist* 112(983): 197–213.

Shannon, C., and W. Weaver. 1949. *The Mathematical Theory of Communication.* Urbana: University of Illinois Press.

19 Modeling Optimal Foraging

Stanton Braude and James Robertson

Introduction

Optimality models, widely used in ecology, are based on the assumption that the strategies of wild organisms are shaped by natural selection so as to maximize gains and minimize expenditure and risk. One of the earliest and clearest uses of optimality models is in comparing foraging strategies. Those individuals that are most efficient in getting food should tend to have higher reproductive success and leave more descendants to future generations, than less efficient foragers.

By formalizing and quantifying models, we can test whether we have identified the important elements of an organism's strategy. We can also determine whether we understand the relationships between these elements. We do not assume that the study species "understands" the model or has a calculator hidden in a cheek pouch or under a wing. But if our model is a good approximation of the important tradeoffs the organism has evolved to balance, then our observations will fit the predictions of the model.

Basic optimal foraging theory was developed by MacArthur and Pianka (1966). Their model predicted that when prey are evenly distributed in space, predators spend less time searching and become more selective in consuming the prey with the highest payoff (in calories). We may also predict that an optimal forager will have a strategy that balances the intake of energy and nutrients against not only the required time but also potential risk.

There are three main categories of foraging models, aimed at answering different questions: prey choice, patch choice, and linear programming. Basic prey choice models assume that the predator has perfect knowledge of its universe and the problem is to maximize caloric gain for search time (affected by prey rareness) and handling time (time to get the good parts from the prey, e.g., clams or lobster have high handling time, berries have little handling time). Basic patch models assume the universe is made of patches of different richness; so the problem also has to include the cost of travel time, and the diminishment of payoff as the predator uses up the prey in a patch (the "marginal returns" diminish). The Pianka and MacArthur model is a classic prey choice model.

Sometimes factors other than calories and time are important. For example, Belovsky (1978) used linear programming to model the optimal moose diet based on constraints of calories, sodium content and rumen capacity. Terrestrial plants offer much better caloric return than aquatics, but only aquatic plants offer sodium—a required nutrient. He found that moose on Isle Royale ate a combination of terrestrial and aquatic plants that matched the predictions of his model, optimizing the salt and calorie combination.

Specific tradeoffs depend on the particular species under study, and its ecological constraints. For some foragers we can make specific predictions about prey choice based on handling time, experience, and encounter rate (e.g., Heinrich, 1979). For others, the marginal value theorem allows us to predict that foragers in a heterogeneous, or patchy, environment will move among patches based on resource depletion in a patch and travel time between patches (e.g., Cowie, 1977).

Brown (1988) also considered tradeoffs in developing the concept of giving up density (GUD) which models when an optimal forager should give up on the current patch and move to another. GUD equals predation risk (P) plus the cost of foraging (C) plus the missed opportunity cost (MOC).

The detailed predictions of any optimal foraging model depend on our understanding of the basic natural history of the study species and how it can best maximize resources, and minimize risk and expenditure of time and energy.

Homework for this exercise takes approximately 2 hours. Don't leave this until the last minute!

Objectives of This Exercise

In this exercise you will apply optimality theory to a novel situation and generate your own testable mathematical model. Your model will predict the behavior of a species you know well, the Halloween trick-or-treater. Your model may include elements of patch choice, prey choice, or linear programming.

Case Study and Data

Every October 31 we are presented with a rare opportunity to observe intensive human foraging behavior. The subjects are children ages 3–12, who scramble from house to house, block to block, and even neighborhood to neighborhood. They are constrained by time and distance. They clearly have preferences for different items. They are driven by their addiction to sugar, by indoctrination with the cultural imperative to acquire and hoard, and by their greedy little natures.

As ecologists we can exploit this annual event. However, field biologists need models and hypotheses to test before we go out into the field to collect data. Your job will be to develop some specific models and predictions.

We have discussed the general principles of optimal foraging theory and a variety of models with predictions of how optimal foragers should respond to differences in patch size, food quality, risk, and competition. Now we want to test the hypothesis that children optimize their foraging behavior during trick-or-treating on Halloween (fig. 19.1).

Your assignment is to develop a mathematical model of how the little monsters (or ghosts, or pirates) should behave as optimal foragers. Your model should make specific predictions about behavior; it should define the elements of the model and the components of your equations. It can be tricky to translate accurately a behavioral prediction into a mathematical model. Some models try to account for all possible factors that may influence strategies, and are, as a result, very complex and difficult to use. Other models include only factors *perceived* to be significantly influential; these may fall short of representing the true nature of the system.

FIGURE 19.1. Little trick-or-treaters like this will be the subject of your foraging model.

Your model should also include specific predictions that can be reasonably tested by simply observing the study subjects. If observations fit the predictions of your model, you can make a strong case that you have correctly identified the important factors that motivate trick-or-treat behavior and the rules that govern decisions of how to forage on Halloween.

You could build a model around the hypothesis that kids are motivated to maximize the amount of chocolate candy collected by optimally visiting different "patches" or neighborhoods. In order to test this general hypothesis you need to make specific predictions. "Natural history" accounts of this "species" preferring chocolate and moving around at night on Halloween would support this general prediction.

Here is a ridiculously exhaustive list of possible parameters. Your model is unlikely to incorporate more than a handful of these, but this should give you an idea of the range of possibilities you can pursue.

Ch_i = number of chocolate bars handed out per house in a given neighborhood (i)

$\overline{Ch}_i$ = average number of chocolate bars handed out per house across all neighborhood

O_i = number of other candy items handed out per house in a given neighborhood (i)

$\overline{O}$ = average number of other candy items handed out per house across all neighborhoods

S_i = number of total candy items handed out per house in a given neighborhood (i)

$\overline{S}$ = average number of total candy items handed out per house across all neighborhoods

$TTBH_i$ = travel time between houses in a given neighborhood (i)

$TTBH$ = average travel time between houses across all neighborhoods

$TTBN$ = travel time between neighborhoods

TT = total time children are allowed outside

TR = portion of TT that children have left until they are called in for supper (that they won't want to eat)

M_i = muggings reported that month in a given neighborhood (i)
$\overline{M}$ = average number of muggings reported that month across all neighborhoods
A_i = assaults reported that month in a given neighborhood (i)
$\overline{A}$ = average number of assaults that month across all neighborhoods
K_i = kidnappings reported that month in a given neighborhood (i)
$\overline{K}$ = average number of kidnappings that month across all neighborhoods   (19.1)

If you hypothesize that risk has a strong influence on trick-or-treater behavior, you could create an index of risk,

RF_i = risk factor for October in a given neighborhood (i)
  = $1(\overline{M_i}) + 3(\overline{A_i}) + 20(\overline{K_i})$
  (This assumes an additive model with the weighting scheme 1, 3, and 20 but your model might assume very different values).   (19.2)

(*Note*: economic models often use the term risk as an indicator of variance [risk of no payoff]. The risk described above is different; it is the risk of getting hurt and is analogous to predation risk. The frantic pace of trick-or-treater behavior may suggest that missed opportunities are an exaggerated fear for these foragers and that the real risks of automobile traffic are undervalued. We could test this prediction, but you must use your own observations to develop your own model.)

If you hypothesize that trick-or-treaters are motivated by something as subjective as happiness, you would have to define what you mean by happiness in your model. An index of happiness, or happiness units, HU, could be

HU_i = happiness units (a measure of the happiness derived from the trick-or-treat experience)
  = $3(Ch_i) + 1(O_i) - (.06)RF_i$ (19.3)
  (This assumes an additive model with the weighting scheme 3, 1, and .06. In another model weighing could be very different. The parameters you choose, how you weigh them and their relationships to one another are all actually hypotheses about the subconscious strategies that govern trick-or-treater behavior.)

A simple model to explain the movements of trick-or-treaters motivated to maximize the number of chocolates collected in any given neighborhood would have to include travel time between houses (TTBH). This model could predict when a child would move to a new neighborhood on the free Halloween shuttle bus provided by the local Dental Association (the bus is free but does not have a set route between neighborhoods). Such a model could be

$$\frac{Ch_i}{TTBH_i} < \frac{\overline{Ch_i}}{\overline{TTBH}}.$$ (19.4)

When this is true, switch neighborhoods.

This model however, does *not* consider the time wasted in moving to a new neighborhood (TTBN) (see fig. 19.2). Please remember to be clear and direct about the assumptions of your model. For example, the models discussed above require that kids have some rough idea of how much chocolate is out there and how far apart houses are in different neighborhoods (maybe older kids know from prior years how much chocolate they and all their friends got and they just need to look down a street to judge the distance between

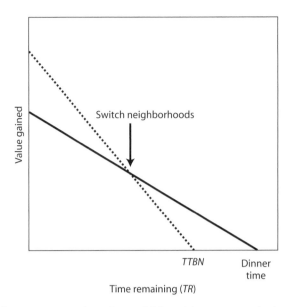

FIGURE 19.2. This figure represents the point at which a trick-or-treater who is optimizing happiness units is predicted to switch neighborhoods. The dashed line represents the happiness gained per unit time for the current neighborhood $(\overline{HU}_i/\,TTBH_i)\times(TR)$ and the solid line represents the happiness gained per unit time for all possible neighborhoods $(\overline{HU}_i/\,TTBH_i)\times(TR)$. Up to the point indicated by the arrow, payoff is better than average and we expect trick-or-treaters to continue in the same neighborhood. Once payoff drops below the average for all neighborhoods, this model predicts that trick-or-treaters will want to switch to a different neighborhood.

houses). However, we need not assume that the kids are consciously making these decisions or calculating the relative payoffs (even if many of them do actually have calculators in their pockets).

In class you will be given real data that may help you make specific predictions with which to test your model. These data have been collected in five different municipalities in the St. Louis metropolitan area (Ladue, Clayton, Brentwood, University City, and the City of St. Louis). To give you some idea of what hypotheses will be testable, note the data we have collected on each municipality.

Basic demographic data will include Total Population, Males, Females, and Median resident age. Data on Land area, Median household income, Median house value will be provided and might be useful with models that involve distances, densities or wealth of residents. For population 25 years and over there will be data on the percentage of the population with: High school or higher, Bachelor's degree or higher, Graduate or professional degree, Unemployed, and Mean travel time to work. For population 15 years and over we will provide family structure data such as: Percent Never married, Now married, Separated, Widowed, Divorced. We will also provide data on single-family new house construction building permits and average costs that might reflect condition of the neighborhoods. Crime statistics will include: murders (per 100,000), rapes (per 100,000), robberies (per 100,000), assaults (per 100,000), burglaries (per 100,000), larceny counts (per 100,000), auto thefts (per 100,000) These data will give you some rough indices of danger, distances to travel between residences, percent of houses likely to have candy, and quality of candy.

Questions to Work on Individually Outside of Class

1. Generate *your own* mathematical model of optimal foraging behavior by trick-or-treaters on Halloween. Define the terms and relationships between the elements of the model. Explain the model in words and then present the mathematical representation and the predictions.

We have presented one model (figure 19.2). Your job is to develop your own model of trick-or-treat behavior.

Ask yourself what really predicts their behavior. Do little kids know which neighborhoods are safe? Do they know which neighborhoods have the best (or most) candy? Do they try to maximize the quality of the candy or just the quantity? Do kids of different ages use the same foraging strategies? You shouldn't try to answer all of these questions with one model. Base your model on what you think is important.

2. Make a graphical representation of your model. Note any points where behavior is predicted to change. Include a descriptive caption under the graph.

It typically requires no more than a page to describe the model and a page for the equations and graphical representation. Your instructor will tell you whether to work alone or with the members of your group outside of class.

Small-Group/In-Class Exercise

Foraging Patterns in Trick-or-Treaters

The university just announced an internal grant program for undergraduate research. We need to draft a grant proposal to study foraging by trick-or-treaters. The proposal must give a clear presentation of the model that you plan to test and its assumptions, and describe exactly how you will test it. (We can worry about the budget and other paperwork another time.)

Your first job is to discuss the different models that the group members came up with and then generate a composite of the best models or the best elements of different models. Spend a few minutes discussing your different models of optimal foraging in children. Discuss the merits and problems with the different models you generated. Then choose the best model(s) in your group. (Do not spend too much time on this job before moving on to the next step.)

Now draft the group proposal.

Introduction: Explain your chosen model and why you think it is a good representation of the foraging behavior of your study species. Be sure to (1) state explicitly the assumptions of the model, (2) outline specific predictions that follow from your model, and (3) present the data that fit the model. Then briefly explain what further data you need to collect next year to finish testing the model.

Preliminary findings: It is an unfortunate fact of modern science that in order to get grants we need to have data supporting our hypotheses before they will give us funds to continue a project. Luckily we have some preliminary data that you can use to test your foraging model. Your instructor will hand out the demographic data we have compiled. Use any of the data on those handouts that are relevant to your model. Use the data for preliminary tests and to make a stronger case for getting the funds that will allow you to do the definitive experiments that you have outlined. (As with any species, there will be complications with experiments on human behavior. However, that does not mean that you cannot test hypotheses with clear experimental design.)

Methodology: Describe a realistic set of experiments that will test whether the patterns of foraging followed by trick-or-treaters fit your model of optimal foraging. Be sure to

- make specific predictions and decision rules
- describe the data that you will need to collect
- describe realistic means of collecting the data

Proposals are judged on both feasibility and clarity. Proposals should always convince the reviewer that your question is interesting; that your experimental design is realistic and will give you unambiguous results; and that you can complete it in the time frame proposed. You also want to convince your reviewer that you are qualified to perform the work. For this assignment, the fact that you were once a child or that you are particularly fond of sweets is less relevant than your ability to get the data you need to test your model.

Use the neighborhood data provided by your instructor. Your written proposal is due at the end of this class period.

References

Belovsky, G. 1978. Diet optimization in a generalist herbivore: the moose. *Theoretical Population Biology* 14: 105–134.

Brown, J. S. 1988. Patch use as an indicator of habitat preference, predation risk, and competition. *Behavioral Ecology and Sociobiology* 22: 37–48.

Cowie, R. 1977. Optimal foraging in Great Tits, *Parus major. Nature* 268: 137–139.

Heinrich, B. 1979. *Bumblebee Economics.* Cambridge, Mass.: Harvard University Press.

MacArthur, R., and E. Pianka, 1966. On optimal use of a patchy environment. *American Naturalist* 100: 603–609.

Section V

Synthetic Exercises and Writing Assignments

20 Evaluating Competing Hypotheses of Regional Biodiversity

Stanton Braude

Introduction and Background

Ecologists work at several scales to understand the underlying causes of species abundance and distribution. This semester you have worked with some of the relevant theories. On the broad geographic scale, there are a number of competing hypotheses that work to explain global differences in biodiversity. In an earlier exercise (chapter 9) you used MacArthur and Wilson's species-area model for examining diversity on islands. On mainland continents, the question remains: why are there more species in the tropics than in temperate regions? The two main competing hypotheses focus on either (1) historical influences or (2) immediate differences in energy and potential for net productivity.

Put simply, the historical hypothesis is that the age and geologic history of continents are the main factors influencing the trend toward greater biodiversity in the tropics. Alternatively, the energy hypothesis suggests that the pattern of global biodiversity arises because there is more available energy (to support more biomass in general) and thus a low rate of extinction in the tropics.

Homework for this exercise is a cooperative effort; your group will need to meet outside of class. Budget approximately 2 hours.

Objectives of This Exercise

- Organize and summarize scientific arguments and data in a coherent and convincing format
- Evaluate contradictory evidence and determine the source of the contradiction
- Find references and sources in the library

Questions to Work on Individually Outside of Class

Your job in this exercise will be to debate one side of the continental biodiversity issue. However, because compelling arguments must address the opposing hypothesis as well, you need to understand both hypotheses. You must support all arguments with examples, tables, or graphs. Unless the data are your own (unlikely), be sure that you cite your sources when you mention others' work.

Box 20.1 Debate tips from the Washington University Debate Coach:

1. Competitive debate is arranged with one team arguing the affirmative of a proposition and the other team arguing the negative. This debate is different in that the two teams will be debating competing hypotheses, so each team bears the burden of proof in arguing for its hypothesis.

2. When making an argument, there are often multiple examples or sub-arguments that that apply to the debate as a whole. Be sure to discriminate among these arguments/examples. For instance, you may be arguing that marijuana should be legal. In support of that main argument you might claim that marijuana (A) has health benefits, (B) is an excellent renewable resource for paper, and (C) would generate tax revenue if legalized.

3. When organizing your materials, it is best to identify first the one or two main arguments you will be making. Your main arguments will be supported by your supporting arguments and/or examples in this sequence:

   I. Main Argument #1
   >   A. supporting argument #1 and evidence
   >   B. supporting argument #2 and evidence

   II. Main Argument #2
   >   A. supporting argument #1 and evidence
   >   B. supporting argument #2 and evidence

   Organizing your work in this fashion makes it easy to present, understand and remember! Of course, you will have more in your outline substructure (additional arguments, examples, etc.).

4. Support your arguments with evidence. Your claims alone will not be enough to persuade the audience. Although you need to cite your sources of information, you cannot get away with just citing an authority who argees with you. You need to explain the data and the line of argument and then note where it comes from.

5. When you are cross-examining the opposing team, you are required to respond to each of their arguments. To do this effectively, you need to anticipate their arguments in your preparation for class. It is also essential for you to take notes while they are speaking, so that you can make a coherent response. Any of their arguments that you fail to refute are considered "wins" by the other team. The key here: silence is concession. If you don't respond, we all assume you agree with their argument.

6. Cross-examination (cx) is not a time to make arguments. It is for clarification of your opponent's arguments, identification of contradictions, and generally a time for you to get your opponent to say something damaging to his/her position. CX is considered binding: you must answer all questions truthfully, even if it may be damaging to your position.

7. Your delivery should be like that for any presentation. You should be clear, organized, use strong eye contact, avoid all verbal clutter (like, um, uh, ya know, etc), and speak in an articulate and conversational manner. Most speakers speak too quickly when they are faced with an audience. Our advice is that team members practice their presentation in front of each other. Figure out before class who will present which part of the presentation and coordinate your transitions between team members. Produce your visual aides (transparencies of outlines, tables and graphs) early, and practice using them and referring to them. The more comfortable you are in front of an audience, the better will be your presentation.

Over the next week you need to read and understand the basic arguments and evidence in this debate and dig up additional relevant information. Then your group needs to meet outside of class *before* the actual debate, to prepare your case. You will be evaluated on your group's presentation, your performance in rebuttal, and your ability to answer questions about these hypotheses. You are each responsible for making the entire group's performance as clear and convincing as possible.

Your instructor will let you know whether you will have access to an overhead projector or an LCD projector for PowerPoint. Your group will have 10 minutes to present your case, and 10 minutes to answer questions from the opposing group. If time allows, you will have additional time to make a rebuttal argument and/or summary statement. This may be your final opportunity to impress your peers with your ability to think clearly and to present an argument coherently. Don't leave them with a less-than-glowing impression of you (see box 20.1).

The Arguments

Historical Hypothesis for Biodiversity: We will present compelling evidence that the trend toward greater biodiversity in the tropics is the result of the age and geologic history of the American, African, and Asian tropics, and not the result of available energy in tropical areas.

Energy Hypothesis for Biodiversity: We will present compelling evidence that the trend toward greater biodiversity in the tropics is the result of the greater availability of energy in the American, African, and Asian tropics, and not the result of the age or geologic history of tropical areas.

The references section below includes articles and books that will give you some background and basic information. (There are dozens of other articles and books in the library that can help you shape your argument and support it; search them out.)

References

Bohm, M., and P. Mayhew. 2005. Historical biogeography and the evolution of the latitudinal gradient of species richness in the Papionini (Primata: Cercopithecidae). *Biological Journal of the Linnean Scoiety* 85: 235–246.

Bonn, A., D. Storch, and K. Gaston. 2004. Structure of the species energy relationship. *Proceedings: Biological Sciences* 271(1549): 1685–1691.

Boyle, B. 1996. Changes in altitudinal and latitudinal gradients in neotropical montane forests. Ph.D. dissertation. St. Louis, Missouri: Washington University.

Chown, S. and K. Gaston. 2000. Areas, cradles and museums: the latitudinal gradient in species richness. *TREE* 15(8): 311–315.

Currie, D. 1991. Energy and large-scale patterns of animal- and plant-species richness. *American Naturalist* 137(1): 27–49.

Hawkins, B., R. Field, H. Cornell, D. Currie, J. Guegan, D. Kaufman, J. Kerr, G. Mittelbach, T. Oberdorf, E. O'Brien, E. Porter, and J. Turner. 2003. Energy, water, and broad-scale geographic patterns of species richness. *Ecology* 84(12): 3105–3117.

Kaspari, M., S. O'Donnell, and J. Kercher. 2000. Energy, density, and constraints to species richness: ant assemblages along a productivity gradient. *American Naturalist* 155(2): 280–293.

Lantham, R. and R. Ricklefs. 1993. Global patterns of tree species richness in moist forests. *Oikos* 67: 325–333.

Macpherson, E., 2002. Large-scale species-richness gradients in the Atlantic Ocean. *Proceedings: Biological Sciences* 269(1501):1715–1720.

Mittelbach, G., D. Schemske, H. Cornell, A. Allen, J. Brown, M. Bush, S. Harrison, A. Hurlbert, N. Knowlton, H. Lessios, C. McCain, A.McCune, L. McDade, M. McPeek, T. Near, T. Price, R. Ricklefs, K. Roy, D. Sax, D. Schluter, J. Sobel, and M. Turelli. 2007. Evolution and the latitudinal diversity gradient: speciation, extinction and biogeography. *Ecology Letters* 10: 315–331.

O'Brien, E. 1998. Water-energy dynamics, climate, and prediction of woody plant species richness: an interim general model. *Journal of Biogeography* 25(2): 379–398.

Owens, J. 1988. On productivity as a predictor of rodent and carnivore diversity. *Ecology* 69: 1161–1165.

Rahbek, C. 1995. The elevational gradient of species richness: a uniform pattern? *Ecogeography* 18(2): 200–205.

Ricklefs, R. E. and G. L. Miller, 1999. Chapter 29. *Ecology*. New York: W. H. Freeman Press.

Ricklefs, R. E. and D. Schluter, 1994. *Species Diversity in Ecological Communities*. Chicago, IL: University of Chicago Press.

Rosenzweig, M. 1995. *Species Diversity in Space and Time*. New York: Cambridge University Press.

Stevens, R. 2006. Historical processes enhance patterns of diversity along latitudinal gradients. *Proceedings of the Royal Society of London, Series B* 273: 2283–2289.

Terborgh, J. 1992. *Diversity and theTtropical Rainforest*. New York, NY: Scientific American Library, W. H. Freeman Press.

Tilman, D. and P. Kareiva. 1997. *Spatial Ecology*. Princeton, N.J.: Princeton University Press.

Turner, J. 2004. Explaining the global biodiversity gradient: energy, area, history, and natural selection. *Basic and Applied Ecology* 5: 435–448.

Wright, D. 1983. Species-energy theory: an extension of species-area theory. *Oikos* 41: 496–50.

21 Preparing and Evaluating Competitive Grant Proposals for Conservation Funding

Stanton Braude

Introduction and Background

The methods and models you have learned this semester have direct application to understanding and solving environmental problems. You have gained quantitative and critical tools that will help you, whether you confront applied ecological problems as a professional in the field or as an educated citizen working in another field.

In basic and applied ecology, we use these critical skills to plan projects and studies aimed at understanding and solving environmental problems. A key step in this process is convincing others that your project is well thought out and worth the effort (and thus worth their money). A well-crafted project proposal is the tool you use to accomplish this task.

In this exercise you will propose a small study or project to deal with a topic or problem that we have not been able to discuss in class. Past students have written proposals on a wide range of topics: a survey of water quality in local streams where students fish, an education project with an area elementary school, a study of the mating behavior of the tree squirrels that live on campus. You should choose a project that you might actually want to do, although you will not be required to complete the project for this course.

Objectives of This Exercise

This exercise will give you practice in organizing a logical and convincing argument in support of a proposed project. You will also gain experience in weighing the merits of a pool of competing proposals and setting conservation priorities.

Case Study and Data

The grant proposal instructions from CRUNCH are presented in box 21.1.

Questions to Work on Individually Outside of Class

Your instructor will tell you the due date for your proposal and whether printed or electronic format is acceptable. Once all the proposals have been received you will review them

---

Box 21.1 The Grant Proposal Instructions from CRUNCH

CRUNCH (Coalition of Radicals United for Nature, Conservation, and Happiness) is dedicated to promoting conservation of biodiversity. The CRUNCH Conservation fund gives grants of up to $25,000 to worthy projects that will further this aim.

Please submit a brief one-page proposal that includes the following:

1. Title
2. Name
3. Title of the ecological or conservation problem you propose to study or help to resolve and why it is important
4. A brief background/history of the species/location/project
5. A specific explanation of what you propose to do
6. A brief budget breakdown (maximum grant is $25,000)

The proposal must be typed and cannot exceed one page in length (no appendices or additional materials will be considered).

---

before class. In class you will serve as a member of the grants committee of CRUNCH. Use Table 21.1 for summarizing your views of the other grant proposals. It will be very useful when you are discussing them in committee.

CRUNCH funds a wide range of proposals and the focus changes from year to year depending on the makeup of the grants committee. Proposals have ranged from environmental education programs and recycling support to wildlife research and habitat restoration. See sample proposals in Boxes 21.2 and 21.3.

Crunch Grant Review Worksheet

Based on your outstanding performance in this course, you have been nominated to serve on the Board of Directors of CRUNCH (Coalition of Radicals United for Nature, Conservation, and Happiness). CRUNCH's charter is pretty brief for a radical organization; it simply states that they promote conservation of biodiversity. The main jobs of the board of directors are to recruit new members, help raise funds, show up at rallies, and decide how to allocate the budget of the Danny Kohl Conservation Fund. This fund has a principal of one million dollars and earns 3% interest, which can be distributed to support worthy and promising conservation efforts. This year a number of worthy proposals have been submitted to CRUNCH (table 21.1). The budget meeting has been scheduled and your instructor will inform you of the date.

As a responsible board member you need to prepare for the budget meeting. All of the proposals submitted for consideration will be available for you to read. Your instructor will inform you of whether they are posted on the course website or elsewhere.

You will want to make copies of the worksheet on the following page so that you can summarize all of the grant proposals under review.

Please also type a brief (one page) argument in favor of the *one* project that you think should have top priority (not the one you submitted). You should discuss it on the basis of its merit, in the context of the goals of CRUNCH, and in comparison to the other proposals.

TABLE 21.1.
Notes on CRUNCH grants reviewed.

Proposal title	$ Requested	Merits	Problems

Box 21.2

1. Title: Restoring Effects of Natural Floods to the Owens River.

2. Applicant: Juan Gaskinade

3. Conservation Problem: The naturalized species of *Tamarix* (family Tamaricaceae) are as a group considered the second worst invasive plant in the United States. Three to ten species of the genus have infested over one million riparian acres in at least 30 states. The *Tamarix* are mostly halophytes that are able to tap water sources up to 30 m deep in dry areas. Their leaves accumulate salt in glands and then are seasonally dropped. The increased salinity of the soil and dense tall growth of the plants create an inhospitable environment for native organisms. Most of the riparian areas infested have had recent alterations in historical flood cycles due to hydroelectric and flood control dams. Similar rivers without disturbance of flood cycles show significantly lower levels of *Tamarix* invasion. Eradication of the *Tamarix* is possible on some scales. Even after the eradication is complete, and regrowth is eliminated annually, the soil of the riparian habitat is still inhospitable to most native organisms due to salt concentrations.

The objective of this project is to restore the effects of natural flooding to an area after Tamarix removal in order to reinstate original soil composition.

4. Background and History: The Owens River in Mono and Inyo Counties (Eastern California) has been the center of water use controversy for over 120 years. The city of Los Angeles has used the Owens Basin as a water resource, channeling Owens River water into a cement-lined canal which parallels the river, substantially lowering the water table by up to 10 m. The purpose of the dam was not flood control, as the flood plain along the river has never contained any buildings or other structures, but diversion for Los Angeles water supply. The current flood plain contains unused sage scrub lands with limited cattle grazing. Since the damming of the headwaters of the river in 1922, flow had been restricted in the historic river bed to minimal levels. By the time it reached Owens Lake 60 miles south of the dam, the river was seasonally dry. As of 1995, a court ruling has implemented more substantial flows in the historic riverbed. As a consequence, the river is now running year round, and water tables are slowly rising. Dam runoff is steady, but the river experiences no seasonal flooding as it did before 1922. As of 1995, the last of the *Tamarix* was manually removed from the entire 60 mile stretch of the river by the California Conservation Corps. Inyo County Water Department personnel have monitored the river since 1995 to remove any clonal regrowth or seedlings of *Tamarix*. Biological inventories have found that since the *Tamarix* removal, biodiversity levels along the river have not increased significantly, and are still much lower than biodiversity levels on neighboring nonobstructed, *Tamarix*-free rivers, and lower than pre-1922 records for the Owens River. We propose that biodiversity levels will not increase until the soil's salt layer caused by *Tamarix* leaf litter is removed, and soil returns to pre-*Tamarix* condition. A similar project reinstating historic flood patterns after *Tamarix* removal on the Ojos River in Arizona increased native biodiversity levels 420%.

5. General Proposal: This project will include analysis of safe levels of water release from the dam, with the purpose of creating small spring floods not to exceed 50% of the flood plain capacity, and quarterly monitoring of biodiversity levels. This project has been readily endorsed by cattle ranchers on the Owens River, who expect to see significant increases

(continued on following page)

of forage material due to seasonal flooding of what are now considered arid, unproductive lands. The Rancher's Co-op has agreed to match CRUNCH funds dollar for dollar and use their funds to start building fencing 20 m back from the river's bank to limit cattle access, further enhancing the native riparian habitat. The fences will not be an obstruction to native deer, which are able to jump cattle fences.

6. Budget: Analysis of dam release schedules requires the purchase of a computer and stat software: $3,200. Travel expenses and lunch for undergraduate interns to continue quarterly monitoring of biodiversity levels for a period of 5 years: $5,000. Total= $8,200.

Box 21.3

1. Title: Community Relations, Attitudes Toward National Parks, and a Farmer's Guide to Naked Mole-Rats.

2. Applicant: Nora Benjamin

3. Conservation Problem: A major challenge of the Kenya Wildlife Service (KWS, the parastatal that runs the parks) is to develop a cooperative relationship with local people who live near the parks. For many years the experience of local people had been that park rangers arrested poachers and protected the buffalo, elephants and leopards that ravaged their crops and livestock. It was difficult for them to see any real benefits to having national parks or to cooperating with the KWS officers.

In 1994 I produced and provided posters about naked mole-rats to the education department of KWS. These have attracted a great deal of positive attention at the education offices in Meru and Langatta. I have now also produced a brief pamphlet about naked mole-rats specifically for Kenyan farmers.

The objective of this project is to provide these pamphlets to local farmers via the KWS education department. Unlike some of the materials that the department distributes, this pamphlet will be of practical interest to local farmers. It is intended that they will not only learn about naked mole-rats, but they will also appreciate that the information came from KWS and that research in the national parks can help them directly.

The educational benefits of this project will include public education about a local wild species. The conservation benefits will be to foster a more positive attitude toward the Kenya Wildlife Service and provide local farmers with a concrete benefit of wildlife research in the parks. This program also encourages an ecologically sound traditional pest management method and discourages use of poisons that can become amplified higher in the food chain (e.g., in owls that eat poisoned mole-rats), or that can persist in the soil.

4. Background and History: I have been working in Meru National Park in Kenya since 1986. My field work there is on the behavior and natural history of the naked mole-rat (*Heterocephalus glaber*). There is high scientific interest in naked mole-rats because they have a social system like honeybees, with a queen and sterile workers. I am the only scientist studying wild naked mole-rats.

In my years of work in Kenya, I have found that local farmers are very curious about this species because it can be a serious garden pest. Because they live in very large colonies (up to 300 individuals), naked mole-rats can devastate a field in a few weeks. While local farmers

(continued on following page)

have very effective traditional methods for trapping naked mole-rats, they are generally not aware of how large the colonies are and limit their trapping to a few animals at a time. Therefore, they rarely alleviate their crop pest problem.

This study focuses on naked mole-rats in eastern Kenya. Naked mole-rats live in underground burrows and feed on roots and tubers. The species range covers the horn of Africa: northeastern Kenya, southeastern Ethiopia, and most of Somalia. In over 90% of this range the mole-rats are in unarable land and thus do not come into contact with people. However, in a small part of their range they enter cultivated land. I propose to distribute *Farmer's Guides* in areas where naked mole-rats are found surrounding the following parks: Meru, Tsavo, Kora, Marsabit, and Samburu.

5. General Proposal: The *Farmer's Guide to the Naked Mole-Rat* has been reviewed by Lucy Muthee (Education Officer, KWS), Connie Maina (Public Relations Officer, KWS) and Wandia Gichuru (Research Officer, KWS) and translated into Kiswahili. The final version will be laser printed in the United States before I return to Kenya in May, 2000. Ten thousand copies of the guide will be printed and folded in Nairobi and donated to the education department of KWS.

6. Budget: A camera-ready original of the farmers guide will be laser printed in St. Louis. 10,000 copies of the guide will be printed and folded in Nairobi and donated to KWS Education Department. At 2.5 Kenya shillings per copy and an exchange rate of 50 Ksh per dollar, the total cost will be $500.

22 Tracing the History of Scientific Ideas: From Darwin, Connell, or Soule to the Present

Bobbi S. Low

Introduction and Background

"There's nothing new under the sun." In research, as in many other arenas, this is true: even the most original of our ideas have precursors, and even your most brilliant idea is not absolutely new. "If I see further, it is because I stand on the shoulders of giants," wrote Isaac Newton to Robert Hooke in the seventeenth century. It is important always to set your work into context, and to give credit to those whose earlier work has helped you see far. Biologists are generally careful about explaining the history and development of their ideas. Not only don't you want to make enemies by failing to give credit to other researchers for their ideas, but also you don't want to look like a naïve novice who doesn't know the literature in his or her own field.

How do we gain scientific knowledge? It is easy to accept it as an almost magical process, if you are only a consumer, not a producer, of information. Look in a textbook: you seem to see fully completed explanations with only supportive evidence, rather than questions about phenomena that lead to more questions and more work. You might come away thinking the world is full of answers, and no real questions remain—but, as we have tried to emphasize in earlier chapters, nothing could be further from the truth! Before you go on, pick a scientific paper you have read recently (it could be for this class, or another), and as you read the rest of this chapter, look to see how the author handled the concepts we raise in the next few paragraphs.

As a scientist, you will want your research to be good and repeatable; you want to be sure you assess each hypothesis, and identify alternative hypotheses. The first several chapters of this text have aimed to help you learn how to accomplish that in different fields: ecology, conservation biology, and population genetics.

However good your work, it will be lost unless you can set it into context, and relate it to the work of others. To contribute to the growth of scientific knowledge (or even to write an insightful review paper), you must be able to track an idea's growth over time, as people publish papers elaborating and developing it. You must know where your contribution comes from; you must show how and why your work agrees with others', and how and why it disagrees.

The number of professional papers published each year is staggering. As you read more and more in the area of your chosen field, you will be struck by repeated references to a half dozen or so individuals. For example, in behavioral ecology, you will see repeated references to Charles Darwin, R. A. Fisher, G. C. Williams, W. D. Hamilton, R. D. Alexander, R. L. Trivers, E. O. Wilson, Richard Dawkins, and John Maynard Smith. Papers on succession may cite Cowles' work on sand dunes or Clements' work on prairies. Papers on niche may be traced back to Elton, and competitive exclusion can be traced back to Gausse.

If you read carefully, you will be surprised how many of our most basic ideas trace back to Darwin, who offered an explanation—still fruitful and fascinating today—of how the "forces of nature" (e.g., constraints of climate, competition in its various forms, and predator/prey relations) help us understand the diversity of nature. Some modern papers even begin with a short quotation to set the stage. This pattern reflects the fact that no idea is ever entirely new; it has a history of development.

It is very important to know how to trace the origin and development of ideas. You are likely to use three kinds of searches: searching back in time, using the literature cited section of a paper you judge to be excellent; searching forward in time from an important paper, using a tool like the Science Citation Index; and searching the ISI Web of Knowledge/Science site for your topic. Each of these may miss some important steps in the development of an idea, but together they should give you a good picture of the development of your idea.

Objectives of This Exercise

In this exercise you will identify important theoretical concepts in ecology, behavioral ecology, or conservation biology. You will then learn to trace an idea both forward and backward in time in the primary literature.

Questions to Work on Individually Outside of Class

Week 1: Your instructor will provide you with a small selection of papers that cover important theoretical topics; they may be overtly theoretical, or empirical work with a clear theoretical basis. If you have another topic or paper that you would like to substitute, you may consult with your instructor. If your instructor allows you this option you can find excellent empirical papers in recent volumes of *Oikos, Oecologia, Behavioral Ecology,* or *Ecology*. Another good starting point could be a recent review in *Trends in Ecology and Evolution*, the *Annual Review of Ecology and Systematics,* or the *Quarterly Review of Biology*. In choosing a paper yourself, try to choose one that not only interests you, but is neither too recent (you would be amazed at the number of papers never cited again, and the first year or so after publication is often barren; you want to be able to trace the paper's impact) nor too old (in which case, if it is important, you will be swamped by references).

a. Once you have chosen a paper, read it carefully. Is it a theoretical paper, or does it contain empirical data that help examine the impact of a theory, or develop a theory further? Write out what you understand to be the paper's *central theoretical concept of concern.*

Examine the Introduction and the Discussion, and consult the paper's Literature Cited or References section. Who is noted as proposing the original version of the theory? Who is cited for which advances of the theory? Organize these in chronological order, and be sure to make notes on major and minor developments, including actual cases of falsification (proving wrong) of the original hypothesis.

b. Next, go the library, or use your university library online, to retrieve each of the important papers you have noted. Your instructor may have a librarian visit class to help you discover effective ways of searching both back and forward in time.

Did the target paper capture the principles effectively in each case? You have a different goal from that of the author of your target paper: is there more in any of the cited papers that you think is important to put in your analysis? The target paper may be rather limited in scope and may have cited only one aspect from these works, but later, other authors may have elaborated on other aspects of the ideas. So read carefully and keep track! And check the Literature Cited sections of the older papers as you find them.

In writing a review such as this, restrict your subject to what you can realistically expect to cover, and keep in mind the availability of references. For example, if you choose a "kin selection" related topic, you will not be able to include every author that has ever studied kin selection, nor would you want to. Think of your target paper as representative of the developments in this theory over time. Remember that Darwin and other early writers may not have used the actual term (such as "kin selection") we use today. Be prepared to do some reading to figure out how terminologies, as well as ideas, have changed over time. Do not rush this part—the thought and effort you invest at this stage are fundamental to the success of your final paper.

c. At this point, you have traced the growth of theory up to the time of your target paper—and, depending on how good that paper is, you may have a really complete, or somewhat spotty, collection of papers. A literature search or review of the available published information on your subject is the next step. Most university library systems have excellent arsenals of search engines at your disposal. Some key search engines are ISI Web of Knowledge/Science, Cambridge Zoological Abstracts, and BIOSIS Abstracts. Typically, you can access these search engines via the main library home page by clicking on a link like "Networked Electronic Resources" (if you have a librarian working with the class, you will have much more specific information).

d. Next, what has happened since your target paper was published? By tracing papers forward in time, you can do two things. First, you can see how the theory has developed since the target paper, and second, by examining the Literature Cited sections of both the older and the more recent papers, you may find important papers that the author of your target article missed. One effective way to trace an idea forward in time is to use the Science Citation Index, which each quarter lists all works cited by papers in a wide array of scientific journals. It is in all major libraries and often available in CD-ROM. You can also go online to the ISI Web of Science, and search the database for authors you have found to be important: in this way, you can find additional papers they have published.

Week 2: Your instructor may have you bring your work in, to discuss with other students who worked on the same target paper, sharing your successes, and learning effective methods from each other. Go over your findings; check out references others found that you did not, and rethink how you want to write your review.

Week 3: Your paper is due next week. Write it in the form of a literature review, as for *Trends in Ecology and Evolution* (*TREE*) or *Quarterly Review of Biology*, or one of the *Annual Reviews* (such reviews exist for, e.g., ecology and systematics, anthropology, physics). You can also think of this as the introduction to a paper in which you (after reviewing the literature) present new advances in this theoretical area.

Begin with an overview of what the paper will discuss, first treating "general" issues and then becoming more specific. The end of the introduction should have your thesis statement, which summarizes the argument or purpose of your entire paper, as well as outlining the general organization of your paper. By tracing the history of the development of a concept from Darwin to today in your introduction, you can then justify new and interesting questions you think should be investigated on this subject.

The actual paper you write need not be very long. Put most of your effort into thinking about the development of the ideas. Your paper should not exceed five double-spaced pages (literature cited excluded from those 5 pages). Box 22.1 contains a brief example tracing the development of the immunocompetence handicap hypothesis. Your paper will be more explanatory, but this should give you an idea of how we examine the development of a theory.

Box 22.1

A brief example of tracing an idea:
The immunocompetence handicap hypothesis

Folstad and Karter (1992) proposed the immunocompetence handicap hypothesis as an extension of Zahavi's (1975) more general handicap hypothesis for the evolution of secondary sexual characteristics. Zahavi argued that expensive or risky secondary sexual characteristics are attractive to mates because they are honest indicators of the quality of a potential mate (e.g., "If he can get away with dragging that big bright tail around and still survive and get through his day, he must be some healthy guy"). Folstand and Karter argued that because testosterone is often the proximate trigger for those bright showy tails (or other secondary sexual characteristics) and testosterone is immunosuppressive, the males with showy tails are advertising that they can stay healthy even with the handicap of a suppressed immune system. While Folstad and Karter's hypothesis offered an explanation for higher parasite loads in males than in females, it has been used to explain both correlation and lack of correlation between testosterone and reduction in indices of immunity. For example, when high-testosterone males have higher indices of immunity or lower parasite loads, the interpretation has been that those males have such high-quality immune systems that they can overcome the immunosuppression of testosterone (Zuk, 1996). On the other hand, when high-testosterone males have higher parasite loads, the interpretation has been that those males are of such high-quality overall that they can display and attract females despite higher infection due to immunosuppression (Weatherhead et al., 1993; Salvador et al., 1996). And if no relationship is found between testosterone and parasite loads, the argument has been that the high-quality males "are reliably signaling their resistance to parasites since they are still able to fend off parasites in the presence of high circulating levels of androgens" (Saino and Moller, 1994).

Wedekind and Folstad (1994) offered a different explanation, namely, that testosterone suppresses the immune system so that essential resources can be allocated instead to produce secondary sexual characteristics such as horns, songs, or stamina in repeatedly performing a display. However, Hillgarth and Wingfield (1997) pointed out that this explanation is unlikely to account for immunosuppression, because the metabolic resources saved by suppressing immunity would be trivial compared to the associated risk of infection. Hillgarth et al. (1997) offered the alternative hypothesis that the immunosuppressive effect of testosterone protects haploid spermatozoa which are antigenic because they are formed long after the development of the immune system. Braude et al (1999) offer the

(continued on following page)

most convincing explanation of all: that testosterone is not immunosuppressive, but only appears to be so; instead, they argued, testosterone triggers immune resources to leave the bloodstream temporarily and migrate to sites where they are poised to fight infections at wounds. Testosterone triggers this to happen during high risk times such as a mating season. This hypothesis also explains why testosterone correlates with low white cell counts in the blood, because white cells have temporarily moved to other tissues.

References

Braude, S., Z. Tang-Martinez, and G. Taylor. 1999. Stress, testosterone, and the immunoredistribution hypothesis. *Behavioral Ecology* 10(3): 345–350.

Folstad, I., and A. Karter. 1992. Parasites, bright males, and the immunocompetence handicap. *American Naturalist* 139(3): 603–622.

Hillgarth, N., and J. Wingfield. 1997. Parasite-mediated sexual selection: endocrine aspects. Pages 78–104 in D. Clayton and J. Moore (eds.), *Host-Parasite Evolution: General Principles and Avian Models*. Oxford: Oxford University Press.

Hillgarth, N., M. Ramenofsky, and J. Wingfield. 1997. Testosterone and sexual selection. *Behavioral Ecology* 8(1): 108–109.

Saino, N., and A. P. Moller. 1994. Secondary sexual characters, parasites and testosterone in the barn swallow, *Hirundo rustica. Animal Behavior* 48: 1325–1333.

Salvador, A., J. Veiga, J. Martin, P. Lopez, M. Abelenda, and M. Puerta. 1996. The cost of producing a sexual signal: Testosterone increases the susceptibility of male lizards to ectoparasitic infestation. *Behavioral Ecology* 7: 145–150.

Weatherhead, P., K. Metz, G. Bennett, and R. Irwin. 1993. Parasite faunas, testosterone and secondary sexual traits in male red-winged blackbirds. *Behavioral Ecology and Sociobiology* 33: 13–23.

Wedekind, C., and I. Folstad. 1994. Adaptive and non-adaptive immunosuppression by sex hormones. *American Naturalist* 143: 936–938.

Zahavi, A. 1975. Mate selection—a selection for a handicap. *Journal of Theoretical Biology* 53:205–214.

Zuk, M. 1996. Disease, endocrine-immune interactions, and sexual selection. *Ecology* 77(4): 1037–1042.

Take the care necessary to make your paper look professional. A neat, well-typed paper is expected in scientific work. With current word-processing, anything else is inexcusable. A sloppy, careless paper suggests sloppy, careless research.

abiotic factors nonliving ecological influences or elements of a niche (e.g., temperature, humidity, pH).

adaptation trait evolved specifically under the influence of natural selection, in response to environmental constraints.

age distribution (age structure) percentage of a population (individuals) in each age category. *Stable age distribution*: the condition in which the proportion of individuals in each age class does not change (although numbers may, as the population grows, for example). *Stationary age distribution*: the condition in which the numbers as well as the proportion of individuals in each age class remain the same: both the population numbers and age structure remain constant.

alarm call a specific call to warn of a predator's presence. Alarm calls are associated with the presence of close kin (e.g., adult female ground squirrels call more when related young are present); however, when species share predators (e.g., giraffes and baboons), it is common for individuals to pay attention to the alarm calls of other species. Alarm calls typically are broad spectrum in frequency: easy to hear, difficult to locate.

algorithm a specific mathematical procedure, or set of rules, for estimating a parameter, choosing between alternatives, or making a decision.

allele at any genetic locus for a diploid organism, an alternative form of the gene. The condition of two identical alleles is described as homozygous (e.g., aa or AA); two different alleles are described as heterozygous (e.g., Aa).

allometric curve a visual representation of the relationship between two traits, the slope of which indicates the degree to which the traits covary in a nonlinear fashion.

allometric equation the mathematical relationship between two traits, indicating the degree to which the traits covary in a nonlinear fashion, as a function of size.

allometry the study of how traits change as a function of body size (e.g., time to maturity increases, in general, with body size).

allopatry the condition in which two taxa exist, without overlapping, in different geographic areas (cf. sympatry).

alpha diversity the first-order diversity that exists within a habitat, i.e. the number of species found in some standardized area such as an acre, a square mile, or some naturally demarcated habitat area.

alternate hypothesis the alternative (or "research") hypothesis, usually denoted H_a or H_1, is typically nonspecific, and states that the null hypothesis is not true.

altricial "helpless" at birth. Among mammals, species in which newborns are hairless, have closed eyes, and cannot move about much are called altricial. Among birds, this condition is typically called nidicolous (cf. precocial, nidifugous).

ancestral taxon a taxonomic group that has given rise to at least one new daughter species via speciation.

apomorphic character state a derived character state.

aposematic coloration coloration that deters predators by advertising poison or foul taste. See also proaposematic, pseudaposematic.

arthropod any of a number of invertebrate groups in the phylum Arthropoda (literally jointed foot), which includes insects, crustaceons (e.g., crabs, lobsters), arachnids (e.g., spiders, mites, scorpions), and myripods (centipedes and millipedes).

autapomorphy unique apomorphies (character states exhibited by only one taxon) and therefore useless in resolving phylogenetic relationships.

beta diversity reflects the change (or turnover) in species composition over small distances, frequently between adjacent yet noticeably distinct habitat types.

bicornuate uterus a uterus with two arms, along which multiple embryos can attach.

biodiversity see alpha diversity, beta diversity, gamma diversity.

biofilters organisms that clean water as they search for food.

biogeography the study of the spatial distribution of species and the various forces that have given rise to that distribution (dispersal, exclusion, history, etc.). Typically considered at a global scale. *Vicariance biogeography* specifically focuses on historical events which have arisen to separate populations, e.g., the fragmentation of populations by the break up of Pangaea.

bioindicator a species whose presence indicates the health of an ecosystem.

biotic living components of an ecosystem, e.g., predators, competitors; cf. abiotic factors such as climate, slope, soil chemistry etc.

bimodal a frequency distribution with two peaks. Bimodality commonly results when data from two different popuations are grouped.

bivariate each sampling unit is associated with two measurements.

bar charts graphical frequency distributions for categorical data—should be differentiated from histograms by leaving spaces between bars.

bipedal walking on two (bi-) legs. Cf. quadripedal, walking on four legs.

bottleneck a time when a population is very small. If the group survives through the bottleneck, the descendant population is likely to have lost alleles (and thus genetic variation) because of inbreeding or genetic drift.

boxplot A graphical display of numeric data that provides information about the data's central location and variability. Most importantly, they give an indication of the symmetry of the data and the presence of possible outliers. The boxplot's "box" illustrates the range of the middle 50% of the observations. The median is the central line in the box, and the first and third quartiles are the outer boundaries of the box. A vertical line, often referred to as a whisker, connects the upper quartile to the most extreme observation within $1.5 \times$ IQR (interquartile range). A similar vertical line is drawn down from the lower quartile. Observations located beyond the vertical lines are indicated with a symbol, usually a star, dot, or line

cache as a verb, to hide away and store a resource (e.g., food). As a noun, the stored material.

carnassials teeth modified to shear meat with their sharp edges and scissorlike overlap. In carnivora the last premolar and first molar are typically carnassials.

carrying capacity symbolized as K, this is, for a particular organism, the maximum number of individuals a particular environment can support without degradation. Although we treat it as a constant, it is clear that, as ecological pressures change, the actual number that is supportable may also change.

central tendency measures of central tendency describe the values around which observations tend to cluster, i.e., the general location of a distribution on the X-axis.

character state the condition of a feature of an organism. For example, "flower color" is a character; "red flowers" is a character state.

chemotaxis directional movement in response to concentration gradients of chemicals; movement may be toward or away from the stimulus.

cladistics a synonym for "phylogenetic systematics," a method for determining the evolutionary relationships of species using shared, derived states of traits, as opposed to using the total number of shared traits or overall similarity.

clade a monophyletic group of species.

cladogram a phylogenetic tree constructed using the principles of cladistics.

clumped distribution a spatial distribution in which individuals are absent from much of the area under consideration, in contrast to uniform or random distributions.

clutch size the number of eggs laid in a single reproductive bout. In mammals, called "litter size."

coefficient of determination R^2, used in regression analyses to describe the relationship between x and y. R^2 is the proportion (or percentage) of the variability in y that can be explained by the linear relationship between x and y.

coefficient of variation CV, expresses the standard deviation as a percentage of the mean.

colonization rate rate at which a new species successfully inhabit patches or islands where they had previously been absent.

competitive exclusion principle also called "Gause's rule." The principle that species with the same niche cannot coexist; one will ultimately drive the other to local extinction.

confidence intervals intervals that cover the population parameter with a specified probability.

continuous data quantitative measurements defined by the property that between any two values there is always another possible value.

control a group of samples that have not been subject to experimental manipulation, against which the manipulated replicates can be compared.

correlation describes how x and y covary. These two variables are interchangeable, and no dependence of y on x is implied.

correlation coefficient The population correlation coefficient, ρ ("rho"), is estimated by r, the sample correlation coefficient, which describes the tightness and direction of the linear association between x and y.

countershading a form of protective coloration in which animals are darker on their upper (dorsal) surface than on their lower (ventral) surface.

covey a small group of individuals (used in ornithology).

Crepuscular active at dawn and dusk.

crypsis (n) cryptic (adj) having the quality of being hidden or difficult to detect.

decision rule used to determine whether we should reject our null hypothesis. We compare the p-value to H_0. We reject H_0 if $p < \alpha$ and do not reject H_0 if $p > \alpha$.

deme local population; an interbreeding group within a larger population.

demography the study of populations, especially patterns of age-specific survivorship and reproduction; analysis of the effects of age structure on population dynamics.

demographic stochasticity random fluctuations in population size.

density dependent changing with population density, e.g., the percentage of aphids killed by an introduced predatory beetle species.

density, giving-up critical point at which the density of food available no longer outweighs the cost of feeding at that site. The benefit of a food patch is primarily determined by the nutrition value and density of food in an area. A high density of food may encourage an animal to take greater risks while feeding. This approach, which measures remaining density of food items, complements the marginal value theorem, which measures giving-up time in patch.

density independent not changing when population density changes, e.g., the percentage of aphids killed by an application of pesticide.

descriptive statistics help us describe and summarize key characteristics of a population. If our sample is representative of the population, our estimates and the distribution of the sample should be similar to the parameters and distribution of the population.

detritus accumulated organic material from dead organisms.

diameter at breast height (DBH) diameter of a tree trunk 4.5 feet above ground, measured from the uphill slope when the tree is growing on a slope.

diastema the gap between the incisors and remaining teeth in some mammals. Humans do not have such a gap, but horses, for example, do.

dioesious having separate sexes rather than spermatogenic and oogenic organs in the same individuals.

directional selection see natural selection.

discrete measurements Quantitative measurements that can only take certain values and may or may not include zero or negative values.

disequilibrium a population is in disequilibrium for a particular locus when the proportions of homozygous and heterozygous individuals do not match the Hardy-Weinberg proportions (which represent the expected proportions based on the overall frequencies of alleles in the population). Disequilibrium forces include immigration, emigration, selection, and drift. Assortative mating can result in disequilibrium in a population.

dioecious organisms that produce fertile male and female reproductive structures on different individuals (cf. monoecious).

diurnal active during the day.

diversity biodiversity, see alpha diversity, beta diversity, gamma diversity.

Dominance-diversity curve an indication of whether biodiversity is evenly distributed across taxa or whether one, or few, taxa account for a disproportionate amount of the total biodiversity.

drift genetic changes in the frequencies of alleles in a population due to sampling error (in drawing gametes form the gene pool to make zygotes), and from chance (nonselective) variation in the survival and reproductive success of individuals. Drift, like mutation, can be a source of non-adaptive evolution.

duplex uterus a uterus with two separate openings to the vagina; may be found in rodents, bats, and marsupials.

ecomorph a phenotype associated with a particular set of ecological parameters. For example, plant desert ecomorphs typically have thick or reduced leaves; arctic mammal ecomorphs typically have white fur.

ecotoxicology the study of the effects of pollution and toxins on ecosystem structure and function.

ectoparasite a parasite that lives on the outside of its host (cf. endoparasite).

effective population size (N_e) the size of a theoretical population, which would lose alleles, lose heterozygosity, or accumulate inbreeding at the same rate as the real population under consideration. Useful in modeling population changes. For example, a population may include very few males; as a result it may accumulate inbreeding at the same rate as a much smaller population.

emigration rate rate at which individuals leave a population (cf. immigration).

endoparasite a parasite that lives within the host.

endemic a species found only in a certain geographic area (e.g., bison are endemic to North America).

equilibrium number of species the number of species on an island as a result of the opposing effects of repeated immigration by species (only some of which is successful), and extinction of existing species. The number of species in a particular taxon is expected to be stable, but the particular species will change over time.

estimates quantitative descriptions of a sample; they are guesses at parameter values. For example, the maximum value in a sample is a guess at the population maximum.

eusociality the social system involving division of labor in a colony resulting in few breeders, as well as nonreproductive "helpers." Eusociality occurs in many bees and wasps, as well as in termites and naked mole-rats.

evolution change in gene frequencies over time; may arise from the action of natural selection (adaptive evolution), or from random processes (e.g., drift, mutation) that affect gene frequencies.

ex situ conservation efforts that take place somewhere other than the natural habitat of the species involved, as in a zoo.

exaptation a trait that evolved under a certain set of selective pressures, but is later used for a different function (e.g., hair is likely to have evolved for insulation but is now used for ornamentation).

exotic in conservation biology, an introduced, or non-native, species.

exponential distributions A class of continuous probability distributions with constant failure rates, such as fish survival times. The exponential is a severely right-skewed distribution where smallest observations are most frequent and larger observations are relatively rare.

extinction rate rate at which species go extinct. Extinction rate is balanced by immigration rate in MacArthur and Wilson's equilibrium model of island biogeography.

extinction vortex a positive feedback loop in which small populations become ever more susceptible to the effects of drift, inbreeding, and random events, ultimately leading to extinction of the population.

fallows (=fallow) land ploughed and tilled, but left unsown for a season to allow it to recuperate.

falsification the test of a scientific idea is to prove it false. If the idea survives such attempts at falsification, it is likely to be true.

fecundity physiological reproductive potential (e.g., as measured by a female's production of eggs); cf. fertility.

fertility the actual production of living offspring.

fitness the expected contribution of an allele (or genotype, or phenotype) to future generations. Fitness is always relative (the contribution of this gene versus others, for example).

frequency-dependent selection occurs when a gene's fitness (or the fitness of the individual carrying that gene) depends on the frequency in the population. For example, mimics have higher fitness in a population when they are rare than in populations in which they are very common.

frequency distribution list of all possible values of the measurement and the frequency or likelihood with which each value is observed. Frequency distributions may be presented as tables or graphs. Sometimes frequency distributions can be described using mathematical functions. For example, the normal distribution represents distributions described by a specific bell-shaped curve.

functional diversity evaluates the complex interactions among food webs, keystone species, and guilds in an area, which provide a measure of richness in functional features, strategies, and spectra.

gamma diversity the total species richness of a large geographic area, such as a continent, reflecting the combined influence of alpha and beta diversity.

gene flow the movement of alleles between demes or populations.

genetic drift random events, typically in small populations, affecting survival and reproduction that change overall frequencies of alleles in a population.

genotype frequency frequency of a particular combination of alleles in a population.

giving up density see density, giving up.

glade a type of relict desert habitat found on southern slopes in the Ozark range. Glade communities include desert-adapted plants and animals (cactus, scorpions, etc.) which were common in the Ozarks during the globally warm period 5,000 years ago.

granivorous eating grain (seeds) as a mainstay in the diet.

GUD giving up density, see density, giving up.

Hardy-Weinberg equilibrium ideal populations (with infinite size, random mating, no mutation, immigration, or emigration) maintain predictable and constant genotype frequencies based on random assortment and recombination of alleles during reproduction, resulting in H-W.

herbivorous eating plant material as the mainstay in the diet.

heterozygous (adj), heterozygosity (n) having two different alleles of the same gene (e.g., *Aa*); cf. homozygous (having two identical alleles at the same locus).

histogram a graphical frequency distribution of numeric data.

home range the total area in which an animal can be found over a defined period of time.

homoplasy sharing of characters due to any process other than common descent (convergence, parallelism, etc.).

hybrid zone a geographic region where two differentiated populations overlap and can interbreed.

hyperdispersed a distribution that approaches evenness across patches or quadrats (cf. clumped or random).

hypothesis a proposal to explain a set of empirical observations; an argument about how things work. A hypothesis generates a set of predictions which are then used to test the hypothesis (if my hypothesis is true, we should see the following). Alternative hypotheses are competing explanations (if A is true, then B cannot be true, and vice versa).

immigration rate the rate at which individuals join a population from other populations.

inbreeding depression an increased rate of birth defects due to deleterious gene combinations or deleterious homozygous recessives resulting from matings between closely related individuals.

inference the reasoning involved in drawing a conclusion or making a logical judgment on the basis of circumstantial evidence and prior conclusions, rather than on the basis of direct observation. Major forms are deduction (reasoning from the general to the particular) and induction (reasoning from particular facts to general principles).

ingroup a focal taxonomic group (the group of interest).

interval data A type of data without a true zero. The "zero" is often arbitrarily assigned and does not indicate absence (e.g., 0 degrees Fahrenheit). Further, many interval scales are circular.

intrinsic reproductive rate (r) the rate of growth of a population in unlimited conditions. It is a function of age-specific survivorship and reproduction. It should not be confused with net reproductive rate (R_0), which is the replacement rate of a population and is the sum of the $l_x m_x$ column in a life table (see chapter 6).

iteroparity a life history parameter involving repeated reproduction in a lifetime (cf. semelparity).

interquartile range the interquartile range (IQR), $Q_3 - Q_1$, is a measure of the spread of the middle 50% of the data and is only minimally influenced by outliers.

interspecific an interaction or comparison between species (e.g., interspecific competition between cattle and kangaroos for food).

intraspecific an interaction or comparison between members of the same species (e.g., gray squirrels hunting for acorns will experience intraspecific competition with other gray squirrels hunting in the same area).

invasive species an exotic species which spreads rapidly and reduces populations of native species or drives them to extinction.

isocline a graphical representation of the population size above or below which a population is expected to grow or decline. Typically represented as the line along which no change occurs.

lamellae layered scales found on the toepads of some lizards, which may enhance gripping strength.

life expectancy at any particular age, the probable remaining expected life of an individual, given the prevailing age-specific survivorship of the particular population.

life history the sum of measures from birth to death: including age-specific fertility, mortality, movement, etc.

logistic equation model of population growth described by a sigmoid (S-shaped) curve.

lognormal distribution a particular right-skewed probability distribution for which the logarithm of the variable is normally distributed.

macroevolution a term introduced by Theodosius Dobzhansky in 1937, referring to evolution at levels higher than the populational. His view was that evolutionary changes could occur at the level of species and above. Recently, the term has been used simply to refer to large-scale change.

marsupial a member of a group of mammals that typically lack a placenta, and whose females usually have a pouch on the abdomen where newborn young are carried; includes kangaroos, opossums, etc.

mating, random a mating system in which every individual has an equal probability of mating with any individual of the opposite sex.

mean the arithmetic mean is calculated by summing all observations and dividing by the number of observations.

median the median is the center of the data when the observations are ordered from smallest to largest.

metapopulation a series of linked subpopulations, having limited migration between them. The dynamics of the metapopulation should be seen as the sum of the dynamics of the individual subpopulations.

microevolution evolutionary changes on the small scale, such as changes in gene frequencies within a population; includes, but is not limited to, adaptation to local environments.

microsatellite simple sequence repeats of nuclear, mitochondrial, or chloroplast DNA. Microsatellites are useful in estimating kinship and population structure because they are typically neutral and rapidly evolving.

mimicry a condition in which a model species, always distasteful or dangerous, may be imitated (physically, behaviorally) by a mimic species. The mimic may be delicious (Batesian mimicry: e.g., monarch and viceroy butterfly mimics), or distasteful (Mullerian mimicry: e.g., the convergence of several species of poisonous dendrobatid frogs to look alike).

minimum viable population (MVP) the smallest population that, when isolated, has a 99% chance of existing for 1,000 years despite demographic stochasticity, genetic drift, environmental changes, and natural catastrophes.

mode the value or category that occurs most frequently in the sample, i.e., the value or category with the largest count.

monogamy a mating system in which the variance in male and female reproductive success is similar. Typically, males and females form stable pair bonds (cf. polygynous and polyandrous).

monoecious plants that produce fertile male and female parts in the same individual (=hermaphroditic).

monophyletic group a group of species that includes one ancestral species and all of its descendants. A nonmonophyletic group is simply one that does not satisfy this criterion.

mutualism interactions between two individuals, in which both benefit (usually refers to individuals of different species).

natural selection the process by which individuals that have heritable traits which enhance survival and/or reproduction leave more descendants, on average, than those who have alternative traits. Over time the population is made of a larger and larger proportion of individuals carrying these traits.

net reproductive rate (R_0) number of daughters produced by a mother in a given interval (usually a generation). It is the sum of the $l_x m_x$ column of a life table; not to be confused with r, the intrinsic reproductive rate (see chapter 6).

niche the sum of the biotic and abiotic parameters and interactions that limit the distribution of a species.

nidicolous birds that hatch at an early stage of development, usually without feathers, with eyes closed, and unable to regulate their own temperature. In other groups, the term "altricial" is used.

nidifugous birds that hatch at an advanced stage of development, fully feathered, with open eyes, able to regulate their temperature, sometimes even able to fly. In other groups, the term "precocial" is used.

nocturnal active at night, as opposed to diurnal (active during the day), crepuscular (active at dawn and dusk), or cathemeral (active at any time of day or night).

normal distribution A family of distributions whose particular locations and shapes vary with their mean μ and standard deviation σ. All normal distributions are unimodal and symmetric, but every bell-shaped distribution is not necessarily normal. The normal distribution has a specific arrangement of observations, or weight, in the tails of the distribution. If data are normally distributed, 68.3% of the observations fall within one standard deviation on either side of the mean, 95.5% within two standard deviations, and 99.7% within three standard deviations.

null hypothesis usually denoted H_0, is a specific statement about the population; it is typically the hypothesis of the status quo or no difference.

omnivorous consuming both plant and animal food.

outgroup any group used in a study that is not included in the in-group.

p-value probability of a test statistic as large as, or larger than, the observed, given H_0 is true. It measures the likelihood of the observed value of the test statistic, or a value more extreme, if H_0 is true.

parameters quantitative descriptions of the aspect of concern of a population. For example, the maximum possible, or largest, value is a population parameter.

parastatal fully or partially state-owned corporation or government agency. Common in LDCs, and centrally planned economies.

parsimonious (adj), parsimony (n) the principle that the simplest explanation, the one that requires the fewest hypotheses, is the one most likely to be correct. In phylogenetic work, refers to a rule used to choose among possible cladograms, which states that the cladogram implying the least number of changes in character states is the best.

phloem a layer of tree tissue just inside the bark that conducts food from the leaves to the stem and roots.

phylogenetics the science of organizing living taxa according to evolutionary relationships and using cladistic methodology to discover those relationships.

phytophagous eating plant material.

pipping when a bird embryo pecks a small hole in the eggshell with a short, pointed, calcareous hardening on the tip of the upper beak.

placental a group of mammals having a special tissue that provides the fetus with oxygen, water, and nutrients from the mother's blood and secretes the hormones necessary for successful pregnancy (the placenta); includes all mammals except monotremes and marsupials.

plankton organisms floating in the water column, unable to move more than short distances under their own power.

plesiomorphic character state an ancestral character state.

plumage a bird's feathers: the light horny waterproof structures forming the external covering of birds.

Poisson distributions model the distributions of *discrete random variables*. Commonly, the Poisson distribution is used when the variable of interest is a count of the number of rare events that occur in a certain time interval or spatial area.

polarity distinguishes between plesiomorphic and apomorphic characters.

polyandry a mating system in which female reproductive success varies more than male success. These systems are rare, because under most ecological conditions females profit from spending more parental effort than males, and less mating effort. Few mammal species are polyandrous (e.g., tamarins); more bird species are (e.g., jacanas, in which a female holds a territory containing several male-guarded nests).

polygyny a mating system in which male reproductive success varies more than female reproductive success. In these polygynous systems, males typically spend little parental effort; the systems may be territorial, lek (symbolic territories), dominance hierarchy, or scramble competition. In all cases, a few males in a population get the majority of matings (leading to high variance in male success) (cf. monogamous, polyandrous).

polymorphism differences in DNA sequences among individuals.

population (ecological) a group of interacting conspecifics; (evolutionary) a group of interbreeding conspecifics; (statistical) the entire collection of basic units and associated measurements about which information is desired.

population "regulation" because no population grows without limits, but typically levels off at some number (with fluctuations), we are interested in the factors that limit population growth. These may be things like predators, or density-dependent limiting factors such as nest sites. It is important to note that population regulation is an emergent phenomenon, rather than a unitary population-level process.

precocial able to move and to feed oneself immediately after birth. In birds this is called nidifugous (cf. altricial, nidicolous).

proaposematic aposematic characters associated with poison or venom.

protrusile capable of being thrust forward, as the tongue.

proximate hypothesis an explanation for how a characteristic of an organism works, rather than why it evolved.

pseudaposematic coloration or trait that mimics aposematic appearance in an individual that is not actually poisonous, venomous, or distasteful.

pseudofeces particles trapped in mucus by mussels and then excreted by the inhalant siphon (cf. true feces, which pass through the digestive tract and are subsequently excreted by the exhalant siphon).

quadrat sampling the method of dividing a study site into a matrix of plots and collecting data from a random subset of these plots.

quartile do just what their name implies: they split the sample into quarters.

radius quadrat sampling the method of collecting data from a set of circular plots which have been randomly designated within a larger study area.

R **squared** see coefficient of determination.

rain shadow the leeward side of a mountain range typically receives low rainfall because air passing over the mountains drops its moisture as pressure drops at higher elevation as the wind hits the mountain.

random neither systematic nor simply haphazard. For example, a random number is one selected from a known set of numbers in such a way that each number in the set has the same probability of occurrence.

randomization an array or distribution that is not significantly different from a Poisson distribution, and is likely to have been generated by chance alone.

ratio data quantitative measurements with a true zero.

refugia areas where special environmental circumstances have enabled a species or community of species to survive after extinction in surrounding areas.

regression the goal of regression is to describe the variability of y in terms of the associated x value. In regression, x and y are not interchangeable; x is the predictor, or explanatory, variable and y is the response variable.

regression analysis the regression line describes how y changes linearly as x changes. The "best fit" line is commonly determined using the least squares method, resulting in a line that minimizes the sum of the squared vertical distances between the line and each observation.

replication in experimental design, the attempt to make multiple, independent, observations of the outcome of a set of circumstances. In the cell cycle, replication is the process of duplicating DNA.

reproductive value a concept from Fisher (1928): given the prevailing age-specific fertility and mortality schedules, how many daughters are expected to be born to a female of age x? (occasionally defined as "offspring" rather than daughters; it is important to know which definition is being used).

rescue effect condition in which, because an island is close to a mainland source population, inmigration of new individuals may maintain even a small population.

riparian habitat found along the banks of rivers or streams.

risk in ecology, risk can also refer to the probability of death or injury. In economic models, risk reflects high variance (e.g., high risk of failure).

saltatory jumping. Kangaroos, for example, are characterized by saltatory locomotion. The term *saltatory evolution* refers to the theory that the evolution of a new species from an older one may occur as a large jump, such as a major repatterning of chromosomes, rather than by gradual accumulation of small steps or mutations.

sample a subset of a population. The number of measurements in the sample, or sample size, is typically denoted by the lower-case letter "*n*."

sample standard deviation (*s*) the positive square root of the sample variance.

sample variance (*s*²) the sum of squared deviations from the mean divided by $(n - 1)$.

scatterplot a graphical technique that uses points to represent bivariate measurements. The points are arrayed in a space defined by axes representing the two variables of interest. Scatterplots illustrate the relationship (or lack of relationship) between two variables, x and y.

semelparity the condition of reproducing only once in a lifetime, e.g., Pacific salmon.

setae (sing. seta) stiff hairs, bristles, or bristlelike processes. On feet or toes these can allow an animal to walk up a vertical wall like Spiderman.

sexual dimorphism differences between males and females in a species, as a result of sexual selection; may refer to morphological, physiological, or behavioral traits.

sexual selection selection specific to one sex. It comprises two distinct processes, with typically different outcomes. Intrasexual selection arises from within-sex competition (e.g., for antler size in male deer); while intersexual selection arises from preferences exerted by the opposite sex (e.g., female choice in many species). Intrasexual selection typically eventually stabilizes (too large antlers are not only expensive, but as deleterious as too-small antlers), but intersexual selection, such as female choice for very elaborate and costly displays by males, can lead to runaway sexual selection.

significance level (α) In hypothesis testing, we preset the probability of a Type I error, α, to a small value (often 0.05 or 0.01). This is called the significance level of the test. The smaller the value, the higher the power.

siphon the organs through which mussels draw in and expel water and extract food.

sister group the most closely related taxon or group of taxa to the focal group.

species-area curve the relationship between the sizes of islands, or patches of habitat in a particular region of the world, and the number of species of a particular taxon present.

species diversity measures the number of species in an area and takes into account sampling effects and species abundance.

species richness a measure of the number of species within an area that assigns equal importance to each species.

species turnover rate although the total number of species in any given taxon is expected to remain stable over time, the particular species can change. The species turnover rate reflects how frequently this occurs.

spread measures of spread or variability describe how similar the data are to one another and inform us of the range of likely values.

steppe large expanses of grass-covered plains; found in Siberia, central North America, and parts of southeast Europe.

stochasticity lacking any discernable order or plan. The bouncing of molecules into one another is stochastic. Similarly, encounters between individuals may be stochastic in some environments.

subdigital on the palm or sole of a finger or toe.

sympatry existing in the same geographic area (cf. allopatry).

synapomorphy derived characters (apomorphies) shared by more than one taxon.

target effect in MacArthur and Wilson's island biogeography model, a large island is likely to have a high rate of successful immigration because it presents a significantly larger target, compared to a smaller island, on which immigrants could land.

taxic diversity a measure of the taxonomic distribution of species. Taxic diversity also highlights species that are evolutionarily isolated but are important in a given system.

taxon a named group of organisms (e.g., birds, hominids, crickets).

taxonomy the science of organizing species into groups.

territoriality a mating system in which a reproductively valuable physical area is defended.

test statistic A value calculated from sample data that measures the difference between the sample data and the population as defined by the null hypothesis. This difference takes into account that sample estimates will not necessarily equal population parameters, even when H_0 is true.

tradeoff giving up something valuable in order to gain something else of equal or greater value (e.g., trading off money for time, space, or goods). In life history theory, tradeoffs represent constrained optimization.

transect sampling the method of collecting data along randomly selected lines which traverse a study area.

Type I error occurs when H_0 is true, but is rejected. The rate, or probability, of a Type I error is denoted by the Greek letter α.

Type II error occurs when H_0 is false, but is not rejected. The rate, or probability, of a Type II error is denoted by β.

ultimate hypothesis a hypothesis that explains why a characteristic evolved, rather than how it works.

uniform distributions sometimes referred to as "rectangular" distributions; they are symmetric but differ from the normal in that each value has the same frequency of occurrence.

unionid bivalve a family of mussels characterized by filter feeding; they inhabit the sandy or sedimentary bottoms of lakes and streams.

xylem the supporting, water-delivering vascular tissue in plants.

Contributors

James Beck is a Postdoctoral Research Associate in the Department of Biology at Duke University. His research focuses on resolving complex evolutionary relationships among poorly differentiated plant lineages.

Stanton Braude is Research Assistant Professor in Biology at the University of Missouri at St. Louis and Senior Lecturer at Washington University in St. Louis. In addition to twenty years of demographic work on naked mole-rats in Kenya, he works on steroid mediated wound healing, rhinoceros population genetics, and canine evolution.

Cawas Engineer is a Postdoctoral Research Associate working on plant biofuels at the University of Massachusetts, Amherst. His research expertise lies in elucidating gene regulatory networks in plants in order to generate environmentally resilient, high yield crops.

John Gaskin is a botanist with the USDA Agricultural Research Service in Sidney, Montana. His main research interests are population structure and origins of invasive plants, and how this information can be applied to biological control.

Luke J. Harmon is an Assistant Professor in the Department of Biological Sciences at the University of Idaho. He is interested in the causes and consequences of adaptive radiation, particularly in island lizards.

Jon Hess is a Molecular Geneticist at the NOAA Northwest Fisheries Science Center in Seattle, Washington. He is currently addressing questions regarding reproductive success and phylogeography and works on a multitude of species including steelhead, rockfishes, invasive tunicates, and naked mole-rats.

Jason J. Kolbe is a Postdoctoral Fellow in the Museum of Vertebrate Zoology at the University of California, Berkeley. His main research areas are the ecology, evolution, and genetics of invasive species, particularly amphibians and reptiles.

Kenneth H. Kozak is Assistant Professor and Curator of Amphibians and Reptiles at the University of Minnesota and Bell Museum of Natural History. His research focuses on molecular ecology, phylogeny, and comparative biology of amphibians and reptiles.

Bobbi S. Low is Professor of Resource Ecology in the School of Natural Resources at the University of Michigan; she is also a Faculty Associate in the Institute of Social Research and the Center for Study of Complex Systems. Her main research areas concern ecological influences on life history, demography and behavior.

Rebecca McGaha is a graduate student in Epidemiology at the University of Texas School of Public Health. She specializes in women's health, and is currently investigating late recurrence of disease in early-stage breast cancer survivors.

James Robertson is a quantitative evolutionary theorist and field biologist, working with xeric species from kangaroo rats to collared lizards. He is an award-winning teacher, currently teaching in the San Francisco Bay area.

Tara Scherer holds degrees in Biology and Medicine from Washington University in St. Louis. She was awarded the Best Medical Student in Emergency Medicine, 2008 by the National Society for Academic Emergency Medicine. Her current research focuses on the usefulness of CT scanning as a geriatric diagnostic tool.

Emily D. Silverman is a biological statistician with the Division of Migratory Bird Management in the U.S. Fish & Wildlife Service. She works on the monitoring, assessment, and management of migratory bird populations, with particular emphasis on North American waterfowl.

Beth L. Sparks-Jackson is a graduate student, research assistant, and intermittent lecturer in the School of Natural Resources at the University of Michigan. Her research focus is aquatic ecology although she dabbles in statistics and landscape ecology.

Anton E. Weisstein is an Assistant Professor of Biology at Truman State University, and Editor of the Biological ESTEEM Collection of online curricular modules for computational biology. His main research areas concern population genetics, bioinformatics, and mathematical biology education.

Index